Living Mirrors

Living Mirrors

Infinity, Unity, and Life in Leibniz's Philosophy

Ohad Nachtomy

Oxford University Press is a department of the University of Oxford. It furthers
the University's objective of excellence in research, scholarship, and education
by publishing worldwide. Oxford is a registered trade mark of Oxford University
Press in the UK and certain other countries.

Published in the United States of America by Oxford University Press
198 Madison Avenue, New York, NY 10016, United States of America.

© Oxford University Press 2019

All rights reserved. No part of this publication may be reproduced, stored in
a retrieval system, or transmitted, in any form or by any means, without the
prior permission in writing of Oxford University Press, or as expressly permitted
by law, by license, or under terms agreed with the appropriate reproduction
rights organization. Inquiries concerning reproduction outside the scope of the
above should be sent to the Rights Department, Oxford University Press, at the
address above.

You must not circulate this work in any other form
and you must impose this same condition on any acquirer.

CIP data is on file at the Library of Congress
ISBN 978-0-19-090732-7

CONTENTS

Acknowledgments vii

Abbreviations ix

Introduction: What does infinity have to do with life? 1

1. Introducing the main characters: A conceptual outline of Leibniz's approach to infinity 14

2. Leibniz in Paris: Between infinite number and infinite being 35

3. Leibniz reads Spinoza: Different senses and different degrees of infinity 63

4. Infinity and unity: Mathematics and metaphysics 80

5. Infinity and life: A sketch of Leibniz's development 99

6. Animate and inanimate things, natural and artificial machines 113

7. Living mirrors and mites: Leibniz and pascal 134

8. Created things as infinite and limited 158

9. Monads at the bottom, monads at the top, monads all over 177

10. Life and force 188

Conclusion: The re-enchantment of nature 201

Bibliography 205

Name Index 213

Terms Index 215

ACKNOWLEDGMENTS

Early versions of some chapters of this book were presented and discussed on diverse occasions and in many institutions. I received helpful comments and advice from many colleagues who are too numerous to record here. While I cannot mention all by name, I want to acknowledge my great intellectual debt to the following individuals: Raphaele Andrault, Richard Arthur, Maria Rosa Antognazza, François Duchesneau, Daniel Garber, Barnaby Hutchins, Michel Fichant, Liat Lavi, Mogens Lærke, Pauline Phemister, Donald Rutherford, Justin Smith, and Reed Winegar.

This book was published with the support of the Israel Science Foundation. I was fortunate to receive several Israel Science Foundation research grants (469/13, 302/16, and 10/19) that allowed me to benefit from the dedicated work of Liat Lavi, Barnaby Hutchins, Fabrizio Baldassari, and Noam Hoffer. Liat went through very early versions of this book, edited the first two chapters, and helped me to bring the manuscript into its present form; Fabrizio Baldassari was very helpful in verifying and correcting the sources; Noam Hoffer provided useful advice and comments; and Barnaby Hutchins edited the whole manuscript twice with great care and intelligence. I am extremely grateful to Barnaby, Fabrizio, Liat, and Noam. I am also grateful to three anonymous referees for OUP whose comments and suggestions had a significant impact on the final form of this book.

In addition, this book benefitted from a fellowship at the Paris Institute for Advanced Studies (France), with the financial support of the French State programme "Investissements d'avenir," managed by the Agence Nationale de la Recherche (ANR-11-LABX-0027-01 Labex RFIEA+).

Some of the chapters in this book draw on and develop material published elsewhere. In particular, chapters 2 and 3 draw on Nachtomy (2011); chapter 4 draws on Nachtomy (2015); chapter 6 is a revised version of Nachtomy (2010); chapter 8 draws on Nachtomy (2016); and chapter 9 on Nachtomy (2018). I would like to thank the journals and the editors for permission to use these works.

ABBREVIATIONS

A	Leibniz, G. W. *Sämtliche Schriften und Briefe*. Hrsg. von der Preußischen (später: Berlin-Brandenburgischen und Göttinger). Darmstadt (später: Leipzig, zuletzt: Berlin): Akademie der Wissenschaften zu Berlin, 1923ff. Citations quote the series, volume, and page. If not otherwise indicated, the reference is to series 6, volume 3.
AG	Leibniz, G. W. *Philosophical Essays*. Edited and translated by D. Garber and R. Ariew. Indianapolis: Hackett, 1989.
AT	Descartes, R. *Œuvres de Descartes*. Edited by C. Adam and P. Tannery. 11 vols. Paris: J. Vrin, 1996.
C	Leibniz, G. W. *Opuscules et Fragments inédits de Leibniz*. Edited by L. Couturat. Paris: Olms, 1961.
Confessio	Leibniz, G. W. *Confessio philosophi: Papers Concerning the Problem of Evil, 1671–1678*. Edited and translated by R. C. Sleigh Jr. New Haven, CT: Yale University Press, 2005.
CSM	Descartes, R. *The Philosophical Writings of Descartes*. Vols. 1 and 2, edited and translated by J. Cottingham, R. Stoothoff, and D. Murdoch. Cambridge: Cambridge University Press, 1985.
CSMK	Descartes, R. *The Philosophical Writings of Descartes*. Vol. 3, edited and translated by J. Cottingham, R. Stoothoff, D. Murdoch, and A. Kenny. Cambridge: Cambridge University Press, 1991.
Curley	Spinoza, B. [Benedict de Spinoza]. *Collected Works*. Vol. 1, edited and translated by E. Curley. Princeton, NJ: Princeton University Press, 1988.
DSR	Leibniz, G. W. *De Summa Rerum: Metaphysical Papers 1675–1676*. Translated and edited by G. H. R. Parkinson. New Haven, CT, and London: Yale University Press, 1992.
E	Spinoza, B. [Benedict de Spinoza]. "Ethics."
EN	Galileo, G. *Opera*. Edizione Nazionale, edited by A. Favaro. Florence: Tip. di G. Barbèra, 1898.

Gebhardt	Spinoza, B. [Benedict de Spinoza]. *Opera*. 4 vols. Edited by C. Gebhardt. Heidelberg: Carl Winter, 1925.
GM	Leibniz, G. W. *Die mathematischen Schriften von G. W. Leibniz*, edited by C. I. Gerhardt. Berlin: A. Asher; Halle: H. W. Schmidt, 1849–63. repr. Ed. Hildesheim: Georg Olms, 1971. 7 Vols.
GP	Leibniz, G. W. *Die Philosophischen Schriften von Leibniz*. 7 vols. Edited by C. I. Gerhardt. Berlin: Weidmann, 1875–1890. [Reprinted Hildesheim: Olms, 1978]
Grua	Leibniz, G. W. *Textes inédits d'après les manuscrits de la Bibliothèque provinciale de Hanovre, publiés et annotés par Gaston Grua, 1948*. 2 vols. Paris: Presses Universitaires de France, 1948.
L	Leibniz, G. W. *Philosophical Papers and Letters*. Edited and translated by L. Loemker. Dordrecht: Reidel, 1969.
Lafuma	Pascal, B. *Œuvres Complètes*. Edited by Louis Lafuma. Paris: Éditions du Seuil, 1963. Cited by section number.
LBr	Leibniz, G. W. *Der Briefwechsel des Gottfrfied Wilhelm Leibniz* [manuscripts] *in der Niedersächsische Landesbibliothek zu Hannover*. Catalogued by E. Bodemann. Hannover, 1889; [Reprint ed. Hildesheim: Olms, 1966]
LDB	*The Leibniz–Des Bosses Correspondence*. Edited and translated by B. C. Look and D. Rutherford. New Haven, CT: Yale University Press, 2007.
LDV	*The Leibniz–De Volder Correspondence. With Selections from the Correspondence Between Leibniz and Johann Bernoulli*. Edited and translated by P. Lodge. New Haven, CT: Yale University Press, 2013.
LLC	Leibniz, G. W. *The Labyrinth of the Continuum: Writings on the Continuum Problem, 1672–1686*. Edited and translated by R. Arthur. New Haven, CT, and London: Yale University Press, 2001.
LR	Leibniz, G. W. *Discours de métaphysique et correspondance avec Arnauld*. Edited by G. Le Roy. Paris: J. Vrin, 1970.
LSC	Leibniz, G. W. *The Leibniz–Stahl Controversy*. Translated and edited by F. Duchesneau and J. E. H. Smith. New Haven, CT, and London: Yale University Press, 2016.
Monadologie	Leibniz, G. W. "The Principles of Philosophy, or, the Monadology" (1714). In AG 213–24.
NE	Leibniz, G. W. *Nouveaux essais sur l'entendement humain*. Translated and edited by P. Remnant and J. Bennett. Cambridge: Cambridge University Press, 1981. [2nd ed. 1996]. Cited by book, chapter, and section. New System Leibniz, G. W. "New System of

	Nature (1695)", English translation in *Philosophical Essays*. Edited and translated by D. Garber and R. Ariew. Indianapolis: Hackett, 1989 [AG], pp. 138–144.
PNG	Leibniz, G. W. "Principles of Nature and Grace, Based on Reason" (1714). In AG 206–13.
Theodicy	Leibniz, G. W. "Essays on the Goodness of God, the Freedom of Man, and the Origin of Evil" (1710). Translated by E. M. Huggard. LaSalle, IL: Open Court, 1985. [First published, London: Routledge & Kegan Paul, 1951]
WFN	*Leibniz's "New System" and Associated Contemporary Texts*. Edited by R. S. Woolhouse and R. Francks. Oxford: Oxford University Press, 1997.
WFP	Leibniz, G. W. *Philosophical Texts*. Edited and translated by R. S. Woolhouse and R. Francks. Oxford Philosophical Texts. Oxford: Oxford University Press, 1998.

INTRODUCTION
WHAT DOES INFINITY HAVE TO DO WITH LIFE?

[F]or everything goes to infinity in nature.
—Leibniz, "Principles of Nature and Grace"[1]

In the "New System of the Nature and Communication of Substance,"[2] published in 1695, Leibniz draws a distinction between natural and artificial machines. A careful look at his motivation and reasoning shows that Leibniz is using this distinction to draw a line between living and nonliving beings; it also shows that, contra Descartes, he takes natural machines to be animate in the sense of having a soul or an active principle. He also takes them to remain machines to the least of their parts—that is, machines that have a structure of machines within machines that develops to infinity.[3] This description of living beings as nested within one another ad infinitum

---

[1] G. W. Leibniz, "Principles on Nature and Grace" (henceforth: PNG), art. 6 (AG 209).
[2] "Système nouveau de la nature et de la communication des substances, aussi bien que de l'union qu'il y a entre l'âme et le corps" ("New System of the Nature and Communication of Substances and of the Union of the Soul and Body"), 1695 (henceforth: "New System").
[3] For example, Leibniz writes: "I define an *organism* or a natural machine, as a machine each of whose parts is a machine, and consequently the subtlety of its artifice extends to infinity, nothing being so small as to be neglected, whereas the parts of our artificial machines are not machines. This is the essential difference between nature and art, which our moderns have not considered sufficiently" (Leibniz to Lady Masham, 1704; GP III:356). "Thus we see that each living body has a dominant entelechy, which in the animal is the soul; but the limbs of this living body are full of other living beings, plants, animals, each of which also has its entelechy or its dominant soul" (*Monadologie* 70).
    See O. Nachtomy, "Leibniz on Nested Individuals," *British Journal for the History of Philosophy* 15, no. 4 (2007b): 709–28. Some of the historical background of this notion is explored in J. Roger, *Les sciences de la vie dans la pensée française au XVIII siècle* (Paris: Armand Colin, 1963); F. Duchesneau, *Les modèles du vivant de Descartes à Leibniz* (Paris: J. Vrin, 1998); A. Pyle, *Malebranche* (London: Routledge, 2003), ch. 7; C. Wilson, *The Invisible World* (Princeton, NJ: Princeton University Press, 1995), ch. 4; J. E. H. Smith, "Leibniz, Microscopy, and the Metaphysics of Composite Substance" (doctoral dissertation, Columbia University, 2000).

is one of the most intriguing features of Leibniz's mature philosophy. The machines within machines are themselves natural machines, of course, and so are also alive. Thus, according to Leibniz, living individuals are organized in a hierarchical structure such that one is nested within another to infinity. And so there is life everywhere in nature.[4]

Versions of Leibniz's notion of nested individuality already appear in his early writings,[5] but are developed more thoroughly in his later writings, especially in his correspondence with Arnauld (1686–87), and some years later in the "New System" (1695), where the notion of a natural machine is introduced.[6] In the "New System," it becomes clear that the nested structure ad infinitum serves to distinguish living beings from nonliving ones. This development is interestingly related to another idea. Shortly after the appearance of the "New System," Leibniz composed a short text in which he responds to Pascal's distinction between the infinitely large and the infinitely small. He writes:

> What he has just said of the double infinity is nothing but an entry point to my system. What wouldn't he have said with his powerful eloquence, if he had gone further, if he had known that all matter is organic, and that the least portion contains, through the actual infinity of its parts, in an infinity of ways, a living mirror expressing the entire infinite universe in an infinity of ways?[7]

That Leibniz defines a living being as "a living mirror expressing the entire infinite universe in an infinity of ways" deserves attention, although as of yet it has received little. For not only does he identify this notion with the only thing that he takes to deserve to be called a substance besides God and monads but he also—and most pertinent to my purposes here—draws an explicit connection between infinity and life. Indeed, it is by virtue of perceiving its own infinite structure that a monad mirrors

---

[4] PNG §4; AG 208.
[5] For example, in his Paris Notes (1672–1676), Leibniz writes: "any part of matter, however small, contains an infinity of creatures, i.e., a world" (A 6.3:478–79). For Leibniz's early views on this point, see C. Mercer, *Leibniz's Metaphysics: Its Origin and Development* (Cambridge: Cambridge University Press, 2001), ch. 7. Mercer argues "that during the winter of 1670–71, Leibniz invented panorganism, according to which the passive principle in a corporeal substance is constituted of a vast collection of corporeal substances, each of which is itself a corporeal substance whose passive principle is so constituted, and so in *infinitum*" (256).
[6] GP IV:482. See also GP VI:544; GP III:340, 356, 565; GP VI:539.
[7] Translated from F. de Buzon, "Double infinité chez Pascal et Monade. Essai de reconstitution des deux états du texte," *Les Études Philosophiques* 4 (2010b): 554.

the infinite universe. Leibniz thus connects Pascal's infinitely large and infinitely small through an isomorphic relation, which he calls "expression."[8]

Leibniz's use of infinity as a defining feature of living beings seems, however, rather surprising. For why would infinity figure in the definition of a living being? At first glance, infinity seems to belong primarily to mathematics, and since human observational capacities are finite and limited, infinity cannot be observed in nature anyway. For these reasons, it would seem that infinity would not have a significant place in an adequate description of nature in general, nor of its living parts in particular.

And yet infinity does play a major role in some views of the natural world. At least one of the crucial transformations in the change from Aristotelian science to the new science has been aptly described as a turn from a closed world to an infinite universe.[9] This turn is usually understood in terms of its undoubtedly significant effect on cosmology and astronomy. Much less attention has been given to the equivalent turn from the finite to the infinite in the life sciences, thanks to the invention of microscopy, whereupon new worlds within worlds of *animalcula*, or minute animals, were discovered.[10]

One might say that, while the invention of the telescope facilitated a view of infinite space at the macroscopic level, the invention of the microscope facilitated a conception of the living world as extending to the infinitely small at the microscopic level. In both cases there is new empirical insight, new phenomena, and a demand for new explanations.[11] Pascal's eloquent remarks regarding the infinite expanses of

---

[8] I discuss these points in detail in chapter 7, this volume.
[9] A. Koyré, *From the Closed World to the Infinite Universe* (Baltimore, MD: John Hopkins University Press, 1957).
[10] See Wilson (1995); and C. Wilson, "Leibniz and the Animalcula," in *Studies in Seventeenth-Century European Philosophy*, ed. M. A. Stewart (Oxford: Oxford University Press, 1997), 153–75; Duchesneau (1998); and J. E. H. Smith, *Divine Machines: Leibniz's Philosophy of Biology* (Princeton, NJ: Princeton University Press, 2011).
[11] It is, of course, an open question how important a role these devices play in these transitions, but there is no doubt that both play a considerable role in the development of astronomy, as well as the life sciences. Leibniz himself noted that one could have anticipated the existence of microscopic animals prior to their empirical discovery just "as Democritus foresaw the imperceptible stars in the Milky Way before the discovery of the telescope" (*Gottfried Wilhelm Leibniz: Philosophische Schriften*, ed. and trans. H. H. Holz, 5 vols. [Darmstadt: Wissenschaftliche Buchgesselschaft, 1985], 2:302, translated in Smith 2011, 150). He also notes, as early as 1671 or 1672, that, since "[m]oral and medical matters . . . are the things which ought to be valued above all," "I value microscopy far more than telescopy" (cited in M. R. Antognazza, *Leibniz: An Intellectual Biography* [Cambridge: Cambridge University Press, 2009], 99). Leibniz's point here is clearly that the practical issues of medicine are more important for the improvement of human life than are merely theoretical issues, such as the squaring of the circle. It is curious to note that Leibniz applied himself quite intensively to both practical and theoretical enterprises.

space and the infinite structure of minute animals might serve as an illustration of the significant role that infinity played in the domain of astronomy and physics, on the one hand, and in the life sciences (or the view of living beings), on the other.[12]

Indeed, this period saw intense preoccupation not only with the infinitely large but also with the infinitely small. Infinitesimals (or infinitely small magnitudes) were used not only in pure mathematics but also in describing and understanding motion (as exemplified in Hobbes's notion of endeavor, or *conatus*); forces (as exemplified both in Leibniz's and Newton's work); and the division of matter to infinity; as well as the nature of living beings. Pascal's example of the mite and the observations of the early microscopists, such as Malpighi, Swammerdam, Leeuwenhoek, and Hooke, brought new vistas of the infinitely minute to the new science.

As noted, the invention of the microscope and the discovery of minute animals were among the most significant novelties of the life sciences in the first half of the seventeenth century. Leibniz was well informed about these discoveries and took a keen interest in creating first-hand contact with the microscopists.[13] For example, when he traveled from Paris back to Germany in the fall of 1676, he went via Holland

---

[12] Pascal writes:

> Qu'est-ce qu'un homme, dans l'infini? Qui le peut comprendre? Mais pour lui présenter un autre prodige aussi étonnant, qu'il recherche dans ce qu'il connaît les choses les plus délicates. Qu'un ciron, par exemple, lui offre dans la petitesse de son corps des parties incomparablement plus petites, des jambes avec des jointures, des veines dans ces jambes, du sang dans ces veines, des humeurs dans ce sang, des gouttes dans ces humeurs, des vapeurs dans ces gouttes. Que divisant encore ces dernières choses il épuise ses forces et ses conceptions et que le dernier objet où il peut arriver soit maintenant celui de notre discours. Il pensera peut-être que c'est là l'extrême (PR 172) petitesse de la nature. Je veux lui faire voir là-dedans un abîme nouveau. Je veux lui peindre non seulement l'univers visible, mais encore tout ce qu'il est capable de concevoir de l'immensité de la nature dans l'enceinte de cet atome imperceptible. Qu'il y voie une infinité de mondes, dont chacun a son firmament, ses planètes, sa terre, en la même proportion que le monde visible; dans cette terre des animaux, et enfin des cirons dans lesquels il retrouvera ce que les premiers ont donné, et trouvant encore dans les autres la même chose sans fin et sans repos, qu'il se perde dans ces merveilles aussi étonnantes par leur petitesse, que les autres par leur étendue. Car qui n'admirera que notre corps, qui tantôt n'était pas perceptible dans l'univers, imperceptible lui-même dans le sein du tout, soit maintenant un colosse, un monde ou plutôt un tout à égard de la dernière petitesse où l'on ne peut arriver? . . . Car enfin qu'est-ce que l'homme dans la nature? Un néant à l'égard de l'infini, un tout à l'égard du néant, un milieu entre rien et tout, Il est infiniment [éloigné] des deux extrêmes; et son être n'est pas moins distant du néant d'où il est tiré que de l'infini où il est englouti. (cited in Buzon 2010b, 549–56, 551–52)

See chapter 7, this volume, for a detailed discussion of Leibniz's response to Pascal.

[13] For a very informative summary of the early stages of microscopy from 1655 to 1720 and Leibniz's engagement with it, see Wilson (2007, 153–75).

in order to visit Spinoza. At the same time, it was probably no less important for him to meet with Leeuwenhoek and Swammerdam and get first-hand information on their new observations.[14]

As François Duchesneau has argued, in the background of Leibniz's view there is a new approach to the study of nature, viz., that of *iatro*-mechanism. This tradition draws on two main sources: a commitment to a mechanistic description of natural things advocated by Descartes, and the new discoveries made through the invention of the microscope by Malpighi in Italy, Leeuwenhoek and Swammerdam in Holland, and Hooke in England. These observations not only revealed a world of minute animals (*animalcula*) within animals that were previously invisible,[15] they also enhanced speculations regarding the issues of generation and the structure of organic beings.

At the same time, it is not at all obvious how these new microscopic discoveries pertain to Leibniz's view of living beings. The early microscopists describe the world of minute animals as seen under the lens of the newly invented microscope. Some of them (e.g., Malpighi) use terms such as *emboîtement* and *enveloppement* to describe what they see, or as part of the thesis of the preformation of animals.[16] As I shall argue, Leibniz is using such figures of speech not so much in an empirical sense—that is, as a description of natural phenomena—but also, in a metaphysical sense, as a way of *defining* the very essence of living things. For Leibniz, living beings are seen as infinite, "divine machines"—machines created (and possibly annihilated) by God alone, which remain perpetually active as long as they live.

In his recent book, Justin Smith (2011) argues that Leibniz's theory of organic body indeed develops out of the *anatomia subtilis* of his predecessors, but that he renders it distinctly Leibnizian, by the introduction of infinity as a way of accounting for the fundamental difference between natural and artificial machines.[17] I fully endorse this

---

[14] See E. Aiton, *Leibniz: A Biography* (Bristol: Adam Hilger, 1985), 67ff.; and Antognazza (2009, 176).
[15] See Wilson (1995).
[16] See F. Duchesneau, *Leibniz, le vivant et l'organisme* (Paris: J. Vrin, 2010), 230–37, 380–81.
[17] Smith writes:

> however much the microstructural strain of *iatro*-mechanism provides a starting point for Leibniz's mature conception of organic body, the German philosopher's conception nonetheless amounts to a radical departure from the earlier subtle-anatomical conception of bodies as consisting in numerous little machines. The crucial difference lies in Leibniz's introduction of *infinity* into his account of the assemblage of machines that make up the body: for him, an organic body is contrasted with a mere machine to the extent that there is literally no lower limit to its mechanical composition. Leibniz's theory of organic body indeed develops out of the *anatomia subtilis* of his predecessors, but Leibniz takes his predecessors' work and renders it, so to speak, distinctly Leibnizian, by means of the introduction of infinity as a way of accounting for what he takes to be

point. At the same time, I would like to highlight a serious discrepancy between the microscopists' successful project of discovering smaller and smaller *animalcula* and Leibniz's claim that one of the defining features of living beings is that their nested structure proceeds to infinity. For Leibniz's claim simply could never be supported by observation, no matter how powerful a microscope one might build. One might suppose, as some *iatro*-mechanists have, that as our means of observation will improve, smaller and smaller animals will be discovered.[18] By extrapolation, one might argue that this could go on indefinitely. Yet the claim that the structure of a living creature is infinite cannot be supported on the basis of observation alone, for the ultimate scope of observation is necessarily finite and limited.[19]

This discrepancy between Leibniz's claims and what microscopical observation is capable of establishing makes it clear that, however exciting, the discoveries enabled through the invention of the microscope are not sufficient for justifying Leibniz's use of infinity in the strong way that he does—that is, as a mark (or as a defining feature) of living beings. Indeed, I will suggest that, in addition to the empirical (observational) source, Leibniz's use of infinity in his theory of living beings draws on (1) a mathematical source; (2) a theological source; and (3) a metaphysical source.

The mathematical source is grounded in Leibniz's early work on the calculus and his subtle way of distinguishing between various kinds of infinity. The theological source is mainly related to Leibniz's motivation to distinguish man-made artifacts from divine machines and thus to the issue of creation and the preformation of living things. Preformation is the theory of human, animal, and plant development that supposes the inner structure of animals to be preformed, either in minuscule form or encoded in "rational seeds"; on this account, all living beings, in their preformed state, already existed at the creation of the world. This approach accounts for the apparent generation of new creatures, arguing that it is a development of their preformed nature—what Leibniz, therefore calls "transformation" rather than "generation," as in the model of a butterfly developing from (or being a transformation of) the silkworm.[20]

> not just a difference of degree of complexity between natural and artificial machines, but rather a fundamental difference in kind. (Smith 2011, 98–99)

[18] In 1678, Justel reports that Huygens has brought some microscopes from Holland with which one could see an infinity of small animals in a drop of water, and that this made people conclude that "everything in nature is animated" (A 1.2:354, cited in Wilson 1997, 158).

[19] For a similar critique, see Jeffrey McDonough's review of Smith (2011) in *Notre Dame Philosophical Reviews* 14, no. 4 (2012):12, at http://ndpr.nd.edu/news/30317-divine-machines-leibniz-and-the-sciences-of-life-2/.

[20] In the "New System," Leibniz notes that the "*transformations* of Swammerdam, Malpighi, and Leeuwenhoek, the best observers of our time, have come to my aid, and made it easier for me to admit that animals and all other organized substances have no beginning, although we

The metaphysical source—whose consequences will be developed throughout the book—is grounded in the traditional connection between infinity, perfection, and being, clearly expressed in the absolute perfection and infinity of God. In viewing God as the most perfect Being, Leibniz concurs with the medieval and early modern traditions. But unlike most thinkers in the period, Leibniz breaks with the tradition in ascribing infinity to created beings as one of their essential features—a point that did not receive due attention from Leibniz scholars. According to Leibniz, created substances are imitations of their Creator, seen as "diminutive God-like creatures." Creatures imitate God in realizing and expressing a particular and unique aspect of the *infinite* essence of God. While God's essence includes all possible ways of being, and is beyond any particular determination, an individual creature is seen as a realization of a particular possibility.

As I have argued in previous work, creation, according to Leibniz, is best understood as God's endowing a possible course (or a program) of action with power to act. Power of action, or what Leibniz calls primitive force, renders a possible individual actual; and in becoming actual and thus active, creatures also become perpetual. Since a creature exists as long as it acts, and its very nature is to act, it would always exist (even if its existence is contingent upon God's choice to create it). Since it is their activity that renders creatures naturally indestructible, it is significant that natural machines are divinely created, and they are endowed with both an infinite program of action and an intrinsic power of action. This very power of action, I shall argue at the end of this book, is no other than what Leibniz means by life, such that the principle of life and primitive force play a similar role in Leibniz's metaphysics.

As pointed out here, there is a gap between the attention the literature has paid to the place of infinity in the mathematical and astronomical facets of the scientific revolution and the attention paid to infinity in the life sciences. The discrepancy is partly due to the attitude toward infinity in the very making of the scientific revolution. For there wasn't merely *hesitation* about using infinity to describe the natural world; there were also powerful arguments against doing so. Lurking in the background, of course, is Aristotle's warning that "nature flees from the infinite" (Generation of Animals 1.1.715b 15) and his argument (in Physics, Book 3) that the notion of actual infinity leads to contradictions (see section 1.2). Indeed, the vast majority of late medieval and early modern philosophers adopted this Aristotelian line.[21] In addition, some of the major critics of infinity in the early modern period voiced strong

---

think they do, and that their apparent generation is only a development, a kind of augmentation" (AG 140; GP IV:480). See also Leibniz to Sophie, November 4, 1696; A 1.13:89–93; GP VII:541–44), where he says: "death is nothing but a change of Theatre."

[21] As Murdoch notes, "[o]f all the points made by Aristotle in his treatment of the infinite in Book III of the *Physics*, undoubtedly the one most often repeated by the medieval philosopher

epistemic and theological concerns. For example, both Descartes and Pascal, who were among the most important mathematicians and philosophers of the time, were vigorously opposed to using infinity in describing the natural world. They argued that, owing to the finite nature of our mind, the infinite remains beyond our grasp and must be regarded as incomprehensible to us. In stressing the contrast between the finite and the infinite, they continued a traditional line of delineating an irreconcilable gap between an infinite creator and finite creatures, suggesting that it would be not only cognitively impossible but also morally and theologically wrong for us to investigate the infinite.[22] For his part, Galileo exposed the paradoxical nature of the infinite from a different angle. He argued that, owing to the paradoxes inflicting infinity, even the most basic relations of the quantitative sciences, such as "larger than," "smaller than," or "equal to," do not hold and therefore cannot be used in the realm of infinity.[23] If Galileo's reasoning were granted, it would follow that infinity cannot be used in an adequate description of nature, for, according to the strictures of the new mechanistic philosophy and science, nature must be described in quantitative terms alone. Thus, the Aristotelian negativity toward infinity seems to be corroborated in several significant respects.

At the same time, when we turn our attention to Leibniz, we see that infinity figures in almost every aspect of his philosophy. Infinity plays a crucial role not only in his views of God, space, number, and possible worlds[24] but also in his views of the natural world and in the resolution of the two "labyrinths" the human mind gets entangled in: "the labyrinth of the continuum" and "the labyrinth of human freedom."[25]

Leibniz is clearly not wary of infinity. Rather, he believes that infinity should be admired and investigated; this is partly because of his conviction that created things,

---

was his denying the possibility of an 'actual infinite' of any sort and admitting only of 'potential infinite'" (J. E. Murdoch, "Infinity and Continuity," in *The Cambridge History of Later Medieval Philosophy*, ed. N. Kretzmann, A. Kenny, and J. Pinborg [New York: Cambridge University Press, 1992], 567–69).

[22] Descartes, *Principles of Philosophy*, part I, arts. 26–27 (CSM I:201–102; AT VIIIA:14–15). See Pascal's *Pensées*, fragment 427 (Lafuma). As Pascal notes, we perceive the infinite but do not understand its nature (Lafuma, fragment 418). Descartes's view will be considered in more detail in chapter 2, and Pascal's in chapter 8, this volume.

[23] In his "Discourses and Mathematical Demonstrations Concerning the Two New Sciences," Galileo writes: "For I believe that these attributes of greatness, smallness, and equality do not befit infinities, about which it cannot be said that one is greater than, smaller than, or equal to one another" (EN 77–78; translated by Arthur in LLC 355). This background is presented in more detail in chapter 2, this volume.

[24] "[T]here is an infinity of possible universes in God's ideas" (*Monadologie* 53; AG 220).

[25] See AG 95; L 264.

which he ends up identifying with living beings, bear the mark of their creator—that is, a mark of perfection and infinity imprinted on their very nature. In his early work on the infinitesimal calculus, initiated during his years in Paris (1672–1676) under the guidance of Huygens, Leibniz discovered a rational method for treating infinity in mathematics. By translating infinitesimal quantities into finite ones, arguing that they can be regarded as variables, smaller or larger than any assignable quantity, he showed how to use infinitesimals in calculations. As Richard Arthur has argued, almost simultaneous with his invention of the differential calculus was his conceptualization of infinitesimals as useful fictions—that is, a reading that justifies using them in mathematics without making ontological or metaphysical commitment regarding their existence.[26] Leibniz's sophisticated approach, both mathematically and philosophically, has also contributed to his confidence in investigating infinity and in applying it outside mathematics.

Indeed, as already noted, infinity plays an important role in almost every facet of Leibniz's metaphysics: he is well known for the thesis that the actual world is one of infinitely many possible worlds. These infinite possible worlds are conceived by God's infinite intellect,[27] and God himself is seen as an *infinite* and most perfect Being (*Ens Perfectissimum*). The actual world consists of infinitely many individual substances, each of which expresses all others, such that it stands in infinite relations to infinitely many others and "exhibits an infinite series of operations" ("Comments on Fardella," AG 102). Likewise, for Leibniz, "[e]ach portion of matter may be conceived as a garden full of plants and as a pond full of fishes. But each branch of every plant, each member of every animal, each drop of its liquids is also some such garden or pond" (*Monadologie* 67).

Thus, far from avoiding infinity, as others recommend, Leibniz seems to take infinity to be indispensable for an adequate description of nature. As he writes to Foucher, "I am so much in favor of actual infinity that, instead of admitting that nature rejects it, as it is vulgarly said, I hold that it affects it everywhere, for better marking the perfections of its author" (Leibniz to Foucher, GP I:416).[28] Leibniz's endorsement of infinity here is crystal clear. Note, however, that it is actual, rather than potential, infinity that he ascribes to nature, and that this ascription is motivated

---

[26] See Arthur (2001, 102–16); R. T. W. Arthur, "Actual Infinitesimals in Leibniz's Early Thought," in *The Philosophy of the Young Leibniz, Studia Leibnitiana Sonderheft*, ed. M. Kulstad, M. Lærke, and D. Snyder (Stuttgart: Franz Steiner Verlag, 2009), 11–28.

[27] *Monadologie* 53; Theodicy §225.

[28] Cf. "Mes méditations fondamentales roulent sur deux choses, savoir sur l'unité et sur l'infini. Les âmes sont des unités et les corps sont des multiplicités, mais infinies tellement que le moindre grain de poussière contient un monde d'un infinités des créatures" (Leibniz to Sophie, November 4, 1696; GP VII:542).

by his theological conviction noted earlier that infinity in nature is the best way of expressing the perfection of its author.

In pointing this out, I wish to draw attention to the fact that Leibniz is careful to employ different notions of infinity in different contexts. He is particularly careful to distinguish between the use of infinity in mathematical contexts, which typically pertain to abstract and ideal entities, and in metaphysical contexts, which typically pertain to real things. An important example of using this distinction can be seen in Leibniz's attempt to resolve the labyrinth of the continuum. The labyrinth is produced, Leibniz argues, because we confuse these realms: the mathematical or the ideal, which is the proper domain of potential infinity, and the metaphysical or the real, which is the proper domain of actual infinity. Hence, the way out of the labyrinth is to disentangle the confusion by carefully distinguishing between these two realms.[29] Leibniz's favorite illustration is that, whereas we can divide a line or any abstract unit in whatever way we want (as ideal things are potentially divisible), real things are already made up in a certain way and must therefore be divided at the (already existing) joints. In other words, real things are already made up in a determinate way and in a sense are already divided.[30]

Both mathematical and metaphysical aspects of infinity played a central role in early modern philosophy. Consider Descartes's (revival of Anselm's) proof for the existence of God—a proof that crucially depends on seeing God as an infinite being.[31] When Leibniz reviews Descartes's proof in 1676, he observes a severe tension between the infinity of God, seen as the subject of all positive attributes or perfections, and the infinity of number, seen as a collection of all units (i.e., categorematically). After engaging with Galileo's *Dialogues*, Leibniz argues that the notion of infinite number, seen as a collection of units, involves a contradiction. Leibniz then seeks to show, I argue, that the notion of an infinite and most perfect Being does not involve a similar contradiction. In other words, Leibniz comes to think that one needs to show

---

[29] "As long as we seek actual parts in the order of possibles and indeterminate parts in aggregates of actual things, we confuse the ideal things with real substances and entangle ourselves in the labyrinth of the continuum and inexplicable contradictions" (Leibniz to De Volder, January 19, 1706; AG 185; GP II:282). This distinction is discussed in more detail in chapter 4, this volume.

[30] I shall attempt to make the scope of "real things" in this context more precise. I will suggest that Leibniz has mainly aggregates (rather than true substances) in mind when he is contrasting the ideal and the real here. For, as we know, substances, for him, are indivisible unities and, strictly speaking, are not made of parts.

[31] Both Descartes and Leibniz use the traditional notion of *Ens Perfectissimum* so that infinity is only implied. But Descartes's analogy in his *Meditations* between God's perfection and human imperfection and God's infinity and human finitude makes the reference to infinity rather clear. In addition, without the supposition that the most perfect Being would involve all perfections (including existence), the traditional proof would not work.

that the notion of an infinite being is possible or self-consistent (something which Descartes, as well as the rest of the tradition, took for granted) because the notion of an infinite number and other infinite magnitudes is contradictory. This problem—how to explicate the conceptual difference between infinite beings and infinite magnitudes—plays a central role in Leibniz's approach to infinity; it is presented in detail in the second chapter.

Interestingly, in their meeting in The Hague (in 1676), Leibniz showed Spinoza his amended version of Descartes's proof of God's existence.[32] A few months before their meeting, Leibniz read Spinoza's letter to L. Meyer—a letter in which Spinoza deals mainly with questions concerning infinity. In this letter, Leibniz finds useful resources in order to account for the difference between an infinite being and an infinite number. Spinoza's approach turns on distinguishing between the different kinds of infinity. Roughly speaking, while he regards the infinity of number as quantitative, he regards the infinity of God as nonquantitative. I present this approach in the third chapter. The distinction between kinds of infinity as, roughly speaking, quantitative and nonquantitative allows Leibniz to account for the difference between the infinity of God, which he regards as possible, and the infinity of number, which he regards as impossible.

As it turns out, Leibniz's response to Spinoza provides evidence not only for his distinction between two kinds of infinity—quantitative and nonquantitative—but also for a third (intermediate) *degree* of infinity. It is worth emphasizing that, in his response to Spinoza, Leibniz draws these distinctions not in terms of kinds of infinity but in terms of degrees.[33]

In chapter 3, I will examine Leibniz's notion of degrees of infinity in the context of his response to Spinoza. At present, let us note that, in his annotations on Spinoza's letter, Leibniz writes: "I set in order of degree: *Omnia; Maximum; Infinitum*." Roughly speaking, between the highest degree of infinity, which Leibniz ascribes to the absolute and necessary being, and the lowest degree of infinity, which Leibniz ascribes to

---

[32] Leibniz traveled back to Germany from Paris via the Netherlands to see Spinoza (among others) in October 1676. See Antognazza (2009, 176).

[33] As he writes in a summary of a conversation with Tschirnhaus on Spinoza's *Ethics*:

> I usually say that there are three degrees of infinity. The lowest is, for the sake of example, like that of the asymptote of the hyperbola; and this I usually call the mere infinite (*tantum infinitum*). It is greater than any assignable, as can also be said of the other degrees. The second is that which is greatest in its own kind (*maximum in suo scilicet genere*), as for example the greatest of all extended things is the whole of space, the greatest of all successives is eternity. The third degree of infinity, and this is the highest degree, is everything (*Omnia*), and this kind of infinite is in God, since he is all one. (February 1676; A 6.3:385; LLC 43)

entia rationis such as numbers and relations, Leibniz invokes an intermediate degree of infinity—a maximum of its kind (*maximum in suo scilicet genere*). It is certainly worth considering what sort of application Leibniz has in mind for this latter degree of infinity.

I will suggest that Leibniz's intermediate degree of infinity can be used productively to characterize the nature of *living* beings, but more generally that of created beings. Unlike the tradition and all major thinkers in his time, who accepted a sharp dichotomy between an infinite creator and finite creatures, for Leibniz, created beings are indeed finite and limited in many respects; but they are also infinite in several important respects. Unlike God, created beings are not absolutely infinite; rather, they are infinite in a sense that each created substance constitutes a maximal development and a determinate way of realizing a unique possibility among all possible individuals—and thus also a unique expression of God's absolute infinity. A created being may be seen as infinite in the sense of being an individual realizing a maximum in its own kind. As Maria Rosa Antognazza recently put this, an interesting consequence of Leibniz's view is that, "[f]rom a metaphysical point of view, each individual constitutes a species" (2009, 57).[34] A central challenge I undertake in this book is to substantiate this suggestion.

I develop this suggestion in chapters 5–8 by examining Leibniz's employment of infinity in the context of his view of living beings. My contention is that the notion of degrees of infinity in general and the intermediate degree of infinity in particular are useful for capturing the way Leibniz is thinking about living beings. It is important to emphasize at the outset that, for Leibniz, living beings are not just a part of nature; rather, they constitute its basic ontology. As Justin Smith recently observed, according to Leibniz, "the world consists in infinitely many eternally existing biological entities. There is, one might say, nothing else."[35] What Smith stresses here by "*biological* entities"—in full awareness of the anachronism of the term—is that Leibniz's fundamental ontology consists of living things and thus ought to be understood against the context of the life sciences of the time.

---

[34] In the "Discourse on Metaphysics" (art. 9), Leibniz writes, "what St Thomas says of angles or intelligences, that any individual is the lowest species, is true of all substances, provided that one takes a specific difference, as the geometers do with regard to their figures" (A 6.4:1541; AG 42). Antognazza makes this comment in relating Leibniz's early work on the principle of the individual (1663) with his later view, as expressed in the "Discourse on Metaphysics" (1686). She does not, however, develop this point. That each individual constitutes a species seems to me to provide an important link between the type of infinity Leibniz ascribes to created substances (as maximum of its kind) and his view of individuality (that individuals are unique by virtue of a complete notion that defines all their predicates—past, present, and future).

[35] In Smith (2011, 6).

At the same time, as we shall see, the question of life is not merely an empirical one; it is, significantly, a metaphysical question as well. For Leibniz ultimately understands the notion of life in terms of animation—that is, the activity of a soul or an entelechy, which is as central for his view of created substance. Hence, the notion of life is part and parcel not only of Leibniz's biology but also of his metaphysics. While the main question I address in this book is how the notion of infinity informs his view of living beings, there is another interesting question in the background. Why does Leibniz use infinity and life so extensively in his metaphysics, and why does their conjunction play such an important role in his philosophy? At the conclusion of the book I will suggest that there is much at stake in the background and that Leibniz attempts nothing less than a re-enchantment of a world that, in his view, had been left disenchanted by Descartes and Spinoza.[36]

---

[36] By using this term (re-enchantment), I do not intend to invoke connotations related to Max Weber's use of Schiller's term to describe the rationalization of modern society, and the narrowing of the scope of rationality to the finding of means for given ends. Rather, I am using the term to describe a view of nature that is opposed to what Leibniz sees as coming from the naturalistic attitude taken up by Descartes and radicalized by Spinoza. In this picture, nature is morally neutral, void of agency, life, and intrinsic ends. Leibniz's re-enchanted view of nature would involve precisely these features: it includes inherent value and norms, agency, life, and built-in ends. While Leibniz himself does not use the term, it seems to me an efficient and evocative way of capturing his deepest concerns about the dangers involved in certain strands of the new philosophy, as well as his own agenda as a response to it.

1

INTRODUCING THE MAIN CHARACTERS
A CONCEPTUAL OUTLINE OF LEIBNIZ'S APPROACH TO INFINITY

1.1 INFINITE NUMBER AND INFINITE BEING

In this chapter, I introduce the central concepts and distinctions Leibniz is using in articulating his view of infinity. In other words, I introduce the main players in this book. These include: Leibniz's rejection of infinite number; his distinction between infinite being and infinite number; degrees of infinity; the distinction between actual to potential infinity; indivisibility; Leibniz's syncategorematic approach to infinite terms; his distinction between infinite number and infinite series; the law of the series; and the distinction between primitive force and derivative force. My aim is to present and clarify at the outset some of the terminology and concepts that I use in order to present Leibniz's approach to infinity. For this reason, I give priority here to conceptual clarity over the details of Leibniz's development. In other words, in this chapter, I draw on Leibniz's texts with no emphasis on the development of his views. In chapters 2 and 3, I give much more attention to presenting Leibniz's views in their precise historical order. Here I seek to clarify the major resources we shall need to present his complex views. At the same time, this also serves as a sketch of (what I take to be) Leibniz's approach to infinity.

1.1.1 Leibniz's Rejection of Infinite Number vis-à-vis Galileo's Paradox

A most natural place—on both historical and conceptual grounds—to begin a presentation of the way Leibniz employs infinity in his philosophy is his clear and consistent rejection of infinite number. From the very beginning of his serious work in mathematics to the very end of his career, Leibniz was very clear in arguing that an infinite number is impossible, or—what amounts to the same thing for him—that the notion of "infinite number" involves a contradiction. As early as 1672, in the *Accessio ad Arithmeticam Infinitorum*, Leibniz writes,

> [a]n infinite number is impossible, [it is] not unity, not totality, but Nothing. Therefore, infinite number = 0.[1]

What Leibniz means by saying that an infinite number is nothing (*nihil*) may not be very clear. It is worth observing therefore that, a year earlier, in 1671, Leibniz defines nothing (*nihil*) as "whatever can be named but cannot be thought" (Antognazza 2009, 95). It turns out that this definition corresponds very well to Leibniz's more elaborate approach regarding the notion of infinite number and some other quantitative notions that he often juxtaposes, such as an infinite line or a greatest shape—all of which he deems to be impossible. He sees such notions as "something that can be named but cannot be thought"—that is, as a concatenation of terms that, upon analysis, turn out to be contradictory and thus refer to no idea at all. On his account, expressions such as "an infinite number" and "the longest line" are made up in *our language* by a composition of terms that, in fact, do not refer to any ideas.

Intuitively, the contradiction in the concept of an infinite number (which he equates with the number of all numbers) arises from the observation that "[e]very number is finite and assignable; [just as] every line is also finite and assignable" (Theodicy §§70, 113). If the very nature of number is a specified sum, or an assignable aggregate of units, the idea of an unlimited number of units makes no sense. As Leibniz puts this in his notes from Paris,

> [T]he last number will always be greater than the number of all numbers. Whence it follows that the number of all numbers is not infinite. Neither, therefore, is the number of unities. Therefore there is no such a thing as infinite number, that is, it is not <possible>. (A 6.3:477; LLC 53. See also A 6.3:463; DSR 7)

For historical and conceptual accuracy, let us note that Leibniz's reasoning is more involved. His rejection of infinite number is informed by his reading of Galileo's *Dialogues on Two Sciences* during 1672–73, and his response to a paradox that Galileo articulates there.[2] In its simplest form, Galileo's paradox arises when we compare two

---

[1] "Numerus infinitus est impossibilis, non unum, non totum, sed Nihil. Ergo Numerus infinitus = 0" (Leibniz to Gallois [*Accessio*], late 1672; A 2.1:349).

[2] "The number of all numbers implies a contradiction, which I show thus: to any number there is a corresponding number equal to its double. Therefore the number of all numbers is not greater than the number of even numbers, i.e. the whole is not greater than its part" (Leibniz to Malebranche, June 22, 1679; GP I:338).

infinite series, such as the series of the natural numbers (1, 2, 3, . . .) with the series of their squares (1, 4, 9, . . .). It is easy to show that each member of the one series can be correlated with a member of the other series, e.g., each natural number with its square: 1 with 1; 2 with 4; 3 with 9; and so on, to infinity, such that the two series would stand in one-to-one relations to one another and thus would seem to be equal in number. At the same time, the series of natural numbers seems to be larger than the series of squares, since it includes members that are not included in the series of squares, (such as 3). Thus, one might say that the series of natural numbers is both larger than and equal to the series of its squares, or that the series of squares is smaller than and equal to the series of natural numbers—a paradox. Galileo's response to this paradox was to claim that the fundamental relations of quantity (smaller than, equal to, and larger than) do not hold in the realm of infinity.

Leibniz's response was to argue that this would violate Archimedes's axiom that the whole is larger than its parts, and therefore infinite number (understood naively) must be regarded as impossible. But, if one is to use infinity in mathematics and philosophy as extensively as Leibniz does, a more nuanced and sophisticated approach to the interpretation of infinite quantities must be developed. This is one of Leibniz's most central motivations for developing a subtle approach to infinity. Indeed, as Richard Arthur has shown, the development of his syncategorematic approach to infinity is one of Leibniz's main achievements in his early years in Paris.[3] This approach is presented in section 1.3.

1.1.2 Infinite Being versus Infinite Number

As I shall argue in more detail in chapter 2, Leibniz's critique of infinite number needs to be situated in a particular historical and philosophical context, viz., as part of Leibniz's approach to possibility and, in particular, his attempt to distinguish between possibilities and impossibilities by providing possibility proofs—what he also calls causal or genetic definitions. In his notes from Paris, Leibniz frequently uses the notion of infinite number (as well as those of infinite line, the most rapid speed, and the largest shape) as an example of an inconsistent notion. Leibniz's concerns about the consistency of the notion of infinite number are at the background of his concerns about the consistency of the notion of the infinite (or most perfect) being. In 1676, we find Leibniz attempting to demonstrate that the notion of the most perfect being is possible. In fact, Leibniz's texts from 1675–76 provide evidence for two related projects: on the one hand, he seeks to show that the notion of an infinite being is possible; on the other hand, he seeks to show that the notion of infinite number (and

---

[3] See Arthur (2009 and 2013).

those of the largest shape and the most rapid speed) are impossible. Since the notions of infinite number and infinite being are analogous in various respects, a question arises whether Leibniz's position (evidenced by this twofold task) is tenable. Since the notion of an infinite being seems to involve infinitely many attributes (much like units in the notion of an infinite number) and since, as we have seen, he regards the notion of an infinite number as impossible, what justifies Leibniz in thinking that the notion of an infinite being is possible? I present this problem in detail in chapter 2 and argue (in chapter 3) that, in order to justify his position, Leibniz distinguishes between different senses of infinity, which he employs in different contexts.

1.1.3 Quantitative vs. Nonquantitative Senses of Infinity

In particular, the most significant distinction Leibniz needs in order to defend his position is between a quantitative and a nonquantitative sense of infinity. The quantitative sense applies to magnitudes and is mainly used in a syncategorematic sense (see section 1.3). The nonquantitative sense of infinity is seen as absolute infinity, which primarily (if not exclusively) applies to God or the most perfect Being. Intuitively, the infinity that the tradition ascribes to God is not one of size or magnitude. But this intuition is not crystal clear, especially if the definition of God is spelled out in terms of his infinite perfections (or attributes), as we find in Descartes, Spinoza, and Leibniz. The comparison between an infinite being and infinite number helps to clarify the sense in which infinity applies to the most perfect being. According to Leibniz, the most perfect being does not admit of part/whole relations (which constitute the sciences of quantity); rather, it has absolute unity and perfection, such that is beyond measure in the sense that it cannot be understood in quantitative terms.

In the *New Essays on Human Understanding*, Leibniz notes explicitly that "The true infinite is only in the *absolute*, which is anterior to any composition and is not formed by the addition of parts" (NE 2.17). Unlike a number, the true infinite is not composed or formed by the addition of parts or units; rather, it is prior to any composition. In this sense, it does not belong to the category of quantity.

It is crucial to observe that Leibniz relates the absolute infinity of God to absolute unity. As he notes in a letter to Des Bosses: "Only absolute and indivisible infinity has a true unity, namely, God" (March 11, 1706; LDB 33). In the same correspondence, he writes:

> I maintain, strictly speaking, that an infinite composed from parts is neither one nor whole, and it is not conceived as a quantity, except through a fiction of the mind. The indivisible infinite alone is one, but it is not a whole; that infinite is God. (September 1, 1706; LDB 53)

For Leibniz, God is not considered a whole because it does not consist of parts and thus does not admit of part/whole relations, which is the fundamental axiom governing the science of quantity. These texts suggest that Leibniz conceives of the infinity of God not in terms of quantity or magnitude of some particular thing, be it as great, large, or long as it may be; rather, he conceives of the infinity of God in terms of absolute perfection.[4] It is, however, far from easy to explicate this sense of infinity in positive terms. In the *Monadologie*, Leibniz provides the following explication:

> God is absolutely perfect—*perfection* being nothing but the magnitude of positive reality considered as such, setting aside the limits or bounds in the things which have it. And here, where there are no bounds, that is, in God, perfection is absolutely infinite. (*Monadologie* 41; AG 218)

The notion of absolute perfection is explicated here as positive reality considered as such, in abstraction of any instantiation. Even though Leibniz is using the term "magnitude" here, it reads more like a measure or degree of positive reality. I think that it cannot be regarded as quantitative and will therefore call it "qualitative infinity." As we shall see, Leibniz is also using the notion of immensity (or immeasurability—in the sense of being beyond measure) to capture God's infinity.[5] But the main point is that the infinity of the most perfect being concerns absolute perfection and unity rather than the infinity of something, be it number, shape, or speed, all of which Leibniz regards as contradictory.

1.1.4 Degrees of Infinity

The qualification of God as absolutely infinite calls to mind Spinoza's distinction between the absolute infinity of God and the infinity in kind of God's attributes.[6] Indeed, even before the *Ethics* was published, Leibniz had the occasion to read Spinoza's letter to L. Meyer (sent in 1663) on the question of the infinite. Luckily,

---

[4] In some texts Leibniz refers to the real infinite as having parts eminently and though perfectly: "the real infinite is perhaps the absolute itself, which is not composed of parts, but comprehends having parts eminently and though perfectly" (GM III/2:500).

[5] "The infinite would be more correctly called the immeasurable (*immensum*)" (A 6.3:475; DSR 27).

[6] Spinoza's definition of God in E ID6 reads, "[b]y God I understand a being absolutely infinite, i.e. a substance consisting of infinity of attributes [*infinitis attributis*], of which each one expresses an eternal and infinite essence. Explication: I say absolutely infinite, not infinite in its own kind; for if something is only infinite in its own kind, we can deny infinite attributes of it. But if something is absolutely infinite, whatever expresses essence and involves no negation pertains to its essence" (Curley 409).

Leibniz not only read and saved Spinoza's letter but also annotated and commented on it. Indeed, I will argue that, in Spinoza's letter, Leibniz could find the gist of a solution to his own problem—namely, a distinction between kinds of infinity (applied respectively to substance, attributes, and modes). Yet, it is curious and significant that Leibniz rephrases Spinoza's threefold distinction between *kinds* of infinity in terms of *degrees*. As he writes: "I set in order of degree: *Omnia; Maximum; Infinitum*."[7] While Leibniz does not often state this threefold distinction explicitly, I will try to show that each of these notions plays an important role in his philosophy.

It is clear that Leibniz ascribes the lowest degree of infinity to mathematical notions or, more generally, to beings of reason (*entia rationis*). It is also clear that he ascribes the highest degree, the absolute sense, of infinity to God. But the intended use of the intermediate degree of infinity—the infinite in kind or maximum—is less clear. One of my central theses in this book is that, between the absolute infinity that Leibniz ascribes to the most perfect being and the lowest degree of infinity that he ascribes to *entia rationis*, he presupposes an intermediate degree of infinity and being, which may apply to created, living beings. Unlike most thinkers in the period, for Leibniz, created beings are also infinite. But their infinity must be distinguished from the absolute infinity of God, and from the mere infinity of series and other beings of reason. Indeed, the infinity adequate for created beings may well be described as an infinity in kind, in distinction from the absolute infinity of God. This claim is developed in chapters 6 and 7. As I will suggest in chapters 8 and 10, this degree of infinity also corresponds to a degree of perfection and being, which is adequate of living things.

---

[7] "I set in order of degree: *Omnia; Maximum; Infinitum*. Whatever contains everything is maximum in entity; just as a space unbounded in every direction is maximum in extension. Likewise, that which contains everything is most infinite, as I am accustomed to call it, or the absolutely infinite. The maximum is everything of its kind (*omnia suis generis*), i.e., that to which nothing can be added, for instance, a line unbounded on both sides, which is also obviously infinite for it contains every length. Finally those things are infinite in the lowest degree whose magnitude is greater than we can expound by an assignable ratio to sensible things, even though there exists something greater than these things" (A 6.3:282; LLC 115).

In a summary of a conversation he had with Tschirnhaus about Spinoza's *Ethics*, he notes: "I usually say that there are three degrees of infinity. The lowest is, for the sake of example, like that of the asymptote of the hyperbola; and this I usually call the mere infinite (*tantum infinitum*). It is greater than any assignable, as can also be said of the other degrees. The second is that which is greatest in its own kind (*maximum in suo scilicet genere*), as for example the greatest of all extended things is the whole of space, the greatest of all successives is eternity. The third degree of infinity, and this is the highest degree, is everything (*Omnia*), and this kind of infinite is in God, since he is all one" (February 1676; A 6.3:385–86; LLC 43).

1.2 ACTUAL INFINITY, POTENTIAL INFINITY, AND INDIVISIBILITY

The distinction between actual infinity and potential infinity goes back to Aristotle's response to Zeno's paradoxes—an array of ancient paradoxes with a similar aim and a similar structure.[8] In one paradox, known as the Dichotomy, Zeno has argued that, in order for a moving body to cover the distance between point A and point B, it would have to cover half of this distance; but, in order to move to the midpoint between A and B, it would have to cover half of that distance, or a quarter of the distance between A and B; and then half of that distance, or one-eighth of the distance between A and B; and so on, *ad infinitum*. Since this would go *ad infinitum*, Zeno argued that a body could not move at all, since its motion could not even get started. Since this argument applies generally to any mutable body, the intended implication is to show that no motion is possible, and that motion is in fact a mere appearance. Zeno's paradoxes generally aimed to show that it is not only motion that is impossible but also that any change at all is impossible, such that in reality there is only a single, unchangeable and indivisible One Being, as Parmenides had suggested.

Similarly, any division would ruin the unity of such a being. Zeno argued that, if something is divisible, it would have to be divided, and further divided, without end, leaving nothing permanently and truly existing. Hence, the conclusion of this variant of Zeno's *reductio ad absurdum* argument is similar: no division at all is possible. Another related paradox concerns the more general notion of plurality. Similarly, Zeno's aim was to reduce the notion of plurality *ad absurdum*. The fundamental intuition here is that, if something is (by its nature) divisible, then this feature (divisibility) must apply to any further division, *ad infinitum*. But, if so, this apparent something would turn out to be nothing at all.[9] As we shall see, a variant of this argument often serves Leibniz to argue (1) against ascribing substantiality to matter (as defined by Descartes—that is, that *res extensa* are supposed to be substantial); and (2) for the postulation of some indivisible elements in order to account for substantiality.

Aristotle responded to Zeno's paradoxes by employing a distinction between potential and actual division to infinity. Aristotle argues that division presupposes potential but not actual parts; the parts of a continuous thing are nothing but

---

[8] For some background, see: http://plato.stanford.edu/entries/paradox-zeno/.

[9] Arthur glosses this argument as follows: "Zeno contended that if there are Many, there must be an assignable number of them. Now, anything that is divisible can always be further divided. Each part could itself be divided into parts, these parts into further parts, and so on without limit. But then there would be no assignable number of parts—a contradiction" (R. T. W. Arthur, *Leibniz* [Cambridge: Polity Press, 2014], 79).

potential parts: they are parts into which it *could* be divided; but they are not actually (or already) divided. Aristotle thus stresses the difference between being divisible and being divided; and he argues that Zeno's paradoxes turn on conflating the two. Aristotle resolves Zeno's paradox as follows. An interval between two points, A and B, is divisible into an infinity of *potential* parts. But since the interval is not actually divided, a body does not have to cover an actual infinity of subintervals in order to move from A to B. The same reasoning applies to numbers, lines, and other magnitudes, in that one could go ahead and divide them further at any given point, but they are not actually divided, so they contain no actual infinity of parts. Rather, the division to infinity is merely potential.[10]

In contrast to Aristotle, Leibniz is a firm advocate of actual infinity. This is clearly seen by passages such as this: "I am so much in favor of actual infinity that, instead of admitting that nature rejects it, as it is vulgarly said, I hold that it affects it everywhere, for better marking the perfections of its author. Thus I believe that there is no part of matter which is not, I do not say divisible but actually divided and, by consequent, the least particle would be considered as a world full of an infinity of diverse creatures" (Leibniz to Foucher, circa 1692–93; GP I:416). In *Monadologie* 65, Leibniz writes: "Every portion of matter is not only divisible to infinity, as the ancients realized, but is actually subdivided without end, every part into smaller parts, each one of which has its own motion" (WFP 277).

Contrast this to Aristotle's view expressed in his *Generation of Animals*: "Nature avoids what is infinite, because infinity lacks completion and finality, whereas this is what nature always seeks" (1.1.715b15). As I argue here later, Leibniz appeals to a sense of infinity that involves precisely absolute perfection and completion. While this sense is exemplified in the absolute perfection of God, there is also a weaker sense of partial perfection in nature—indeed, one that typifies any created substance. This is not so foreign to Aristotle's teleological sense, since Leibniz allows for an increase of perfection in nature such that created things alter toward a higher degree of perfection.

While Leibniz certainly endorses actual infinity, it is important to observe that his position is much more subtle and nuanced than a sweeping endorsement of actual infinity. For he is not endorsing actual infinity universally; rather, his commitment to actual infinity is usually expressed in a particular context, viz., the division of matter to infinity. In the context of ideal, abstract things, however, Leibniz endorses only potential (rather than actual) infinity. Furthermore, he argues that the whole labyrinth of the continuum is due to a confusion of these two realms (the ideal and the real) and the sense of infinity typical to each one of them.

---

[10] See Aristotle's *Physics* 207b.

Leibniz's commitment to actual infinity is already apparent in his early response to Descartes's position concerning the indefinite division of matter, and in his rejection of Descartes's distinction between the indefinite and the infinite.[11] As we shall explore in the next chapter, Descartes has argued that the only thing that deserves to be called "infinite" is God. All other magnitudes that seem to have no limits should be called "indefinite." Leibniz's early response to this attitude can be seen in a list of predemonstrable foundations that he puts together in his very early "Theory of Abstract Motion" (1670–71):

(1) *There are actually parts in the continuum* ... and
(2) *these are actually infinite*, for Descartes' indefinite is not in the thing but in the thinker.
(3) *There is no minimum in space or body*, that is, nothing which has no magnitude or part. (A 6.2:264 N41; LLC 339)[12]

The most curious and intriguing aspect of Leibniz's commitment to the actual infinity of parts in the continuum is that it comes with a rejection of minima (number 3 on the list just cited) and a firm commitment to indivisibles. Indeed, the next item on this list is:

(4) There are indivisible or unextended things (A 6.2:264 N41; LLC 339).

Keeping these commitments in mind, one can gain an important insight into a line of reasoning that cuts deep into the heart of Leibniz's metaphysics: since matter is considered infinitely divisible, and this division is seen as actual, then if anything real or substantial is to exist, one has to postulate some indivisible elements in order to account for such reality. If anything divisible is also destructible, there must be some indivisibles that would prevent, as it were, the disintegration of a body into nothing. In one passage from Leibniz's Paris notes, the character of this *reductio* argument is particularly clear:

> Whatever is divisible, whatever is divided, is altered—or rather, is destroyed. Matter is divisible, therefore it is destructible, for whatever is divided is destroyed. Whatever is divided into minima is annihilated; but that is impossible. (A 6.3:392; DSR 45)

---

[11] See Descartes's *Principles of Philosophy*, part II, arts. 33, 34 (CSM I:237–39; AT VIIIA:59).
[12] See also A 6.1:169 for an earlier endorsement of actual infinity in *de Arte Combinatoria* (1666).

What we see here is that the infinite divisibility of matter constitutes a reason for Leibniz to postulate the existence of some indivisible things. While the nature of what he takes these indivisibles to be undergoes significant changes over Leibniz's career, the structure of his argument remains more or less the same. In his early years, Leibniz considered various alternatives related to his account of infinitesimals, such as points and momentary minds. But he soon realized that only something that is nonextended (and immaterial) would be immune to the destructive danger of divisibility. At the same time, mere mathematical points are indivisible but do not seem able to provide anything substantial; Leibniz would ultimately gravitate toward an account of substantiality in terms of activity and force, which do not admit of (or require) any extension. As we shall see in more detail in chapter 5, Leibniz's commitment to a soul-like principle, seen as a source of life and unity, is connected with this line of inquiry—that is, looking for something that is not subject to division and destruction and yet can ground substantiality. As Leibniz would argue, the only thing that can do that is the (nonextended) reality of action and force. Indeed, this is expressed in his characterization of souls as simple substances or unities that, unlike bodies, are not formed by the composition of parts. As Leibniz would put the point in his famous formulation: "Souls are unities, bodies are multitudes."[13]

In Leibniz's later writings, these indivisible sources of life and activity come to be called "monads." But his commitment to actual infinity remains firm:

> Since monads or principles of substantial unity are everywhere in matter, it follows from this that there is also an actual infinity, for there is no part, or part of part, that does not contain monads. (Leibniz to Des Bosses, March 2, 1706; LDB 25)

It is also important to observe that Leibniz's commitment to the actual division of matter to infinity goes hand in hand with his commitment to a merely potential divisibility of mathematical or abstract things. While a real thing is actually divided *ad infinitum*, a mathematical thing, such as a line or number is only potentially divisible and it has as many potential parts as one would like to produce by dividing it however one would like. Leibniz relates this to his distinction between the ideal and the real,[14]

---

[13] As he writes to Sophie: "Mes méditations fondamentales roulent sur deux choses, savoir sur l'unité et sur l'infini. Les âmes sont des unités et les corps sont des multiplicités, mais infinies tellement que le moindre grain de poussière contient un monde d'un infinités des créatures" (November 4, 1696; GP VII:542). See also Leibniz to Sophie, February 6, 1706; GP VII:566–70; G. W. Leibniz, *The Shorter Leibniz Texts*, ed. and trans. L. Strickland (London and New York: Continuum, 2006), 82.

[14] "In the ideal or continuous the whole is prior to the parts, as the arithmetical unit is prior to the fractions that divide it, which can be assigned arbitrarily, the parts being only potential;

and argues that confusing these two realms produces the labyrinth of the continuum. As he writes,

> As long as we seek actual parts in the order of possibles and indeterminate parts in aggregates of actual things, we confuse the ideal things with real substances and entangle ourselves in the labyrinth of the continuum and inexplicable contradictions. (Leibniz to De Volder, January 19, 1706; AG 185; GP II:282).

The way out of this formidable labyrinth is to keep the real and the ideal distinct and to note that one domain (the real) admits of actual infinity of parts, while the other (the ideal) admits only of potential infinity and potential parts.[15]

1.3 A SYNCATEGOREMATIC APPROACH TO INFINITE TERMS—SMALL AND LARGE

In his late writings, Leibniz refers to his own approach to infinity by using the scholastic distinction between a categorematic and a syncategorematic use of the term "infinity."[16] Roughly speaking, if a term is used categorematically, it signifies a determinate thing or, as we might say today, it has a determinate reference; if a term is used syncategorematically, it does not (by itself) refer to a determinate thing or to a certain entity.[17] For example, terms such as "it" and "some" are commonly used in meaningful sentences, but in themselves they do not refer to determinate things.

---

but in real things . . . the parts are actual, are before the whole" (Leibniz to Nicolas Remond, 1714; GP III:622).

[15] At the same time, I suggest that the scope of "real things" that Leibniz has in mind here may be restricted to aggregates. Most commentators simply take it for granted that by "real," Leibniz is referring to substances, rather than only to collections of them or aggregates. But, on my reading, Leibniz has "aggregates of actual things," and not true substances, in mind in this context. The main reason for holding this restricted reading of "the real" in this context is quite straightforward: if, as we have seen, substances are indivisible unities for Leibniz, then, strictly speaking, they have no parts at all. It goes without saying that, if a substance is indivisible, as Leibniz clearly holds, it cannot be actually divided. Hence, it seems, Leibniz has collections of substances (rather than substances) in mind in this context. For example, Leibniz surely considers the I or the Self, which is his favorite example of a substance, to be real and indivisible.

[16] As Look and Rutherford clarify, "in medieval logic, terms were divided into categorematic, which signify in their own right; and the syncategorematic, which signify only if combined with other terms. Examples of the latter include prepositions, conjugations, and signs of quantity" (LDB 406n5). See also Murdoch (1992, 567–68).

[17] See M. R. Antognazza, "The Hypercategorematic Infinite," *Leibniz Review* 25 (2015): 5.

This points to the grammatical origin of the distinction. Indeed, this seems to be the point that Leibniz seeks to highlight by applying the terms to the usage of "infinite." If the term "infinity" is used syncategorematically, it would avoid the implication that it signifies a genuine whole of some determinate quantity and would thus avoid the contradiction of an infinite number (number being obviously a determinate quantity).

Thus in the "New Essays," Leibniz writes,

> It is perfectly correct to say that there is an infinity of things, i.e., that there are always more than one can specify. But it is easy to demonstrate that there is no infinite number, nor any infinite line or other infinite quantity, if these are taken to be genuine wholes. The Scholastics were taking that view, or should have been doing so, when they allowed a 'syncategorematic' infinite, as they called it, but not a "categorematic" one. (NE 2.17.1, p. 157)[18]

Note that Leibniz draws an analogy here between the infinitely large and infinitely small, and argues that both should be used syncategorematically, such that there is no infinitely large number and no infinitesimal term; rather, one can go as small (or as large) as one needs for the sake of calculation:

> Speaking philosophically, I no more support infinitely small magnitudes than infinitely large ones, or no more infinitesimals than infinituples. For I consider both to be fictions of the mind, due to abbreviated ways of speaking, suitable for calculation, as imaginary roots in algebra are also.[19]

The idea that both infinitely small and infinitely large magnitudes are fictions—useful fictions, that is—is part and parcel of the syncategorematic interpretation of infinite terms. Rather than taking the infinite to be an assignable fixed number or magnitude, the term "infinite" is seen as a variable. It is not used to designate a collection of units, or any determinate thing. In seeing the use of "infinite" as unassignable and unspecified, Leibniz seeks to avoid the inconsistent notion of "an infinite number" for it is of the essence of any number to be finite and assignable (see citations that

---

[18] Leibniz of course does not attempt to remain faithful to the scholastic distinction. As he stresses here, this is the view the scholastics *should* have taken. In a letter to Des Bosses, Leibniz notes: "There is a syncategorematic infinite or passive power of having parts, namely, the possibility of further progress by dividing, multiplying, subtracting, or adding. . . . But there is not a categorematic infinite or one actually having infinite parts formally" (Leibniz to Des Boesses, September 1, 1706; LDB 53).

[19] Leibniz to Des Bosses, March 11, 1706; LDB 33.

follow). The point is not that the term "infinite" refers to an unassignable *number* but, rather, that it does not refer at all—that it does not designate any whole or fixed quantity. Further, this syncategorematic approach seems to hold generally to any infinite quantity, large or small, numerical or otherwise. As Leibniz puts this in the *Theodicy*,

> Every number is finite and assignable; every line is also finite and assignable. Infinities and infinitely small only signify magnitude which one can take as big or as small as one wishes, in order to show that the error is smaller than the one that has been assigned. (Theodicy §70)

He makes a similar point in a letter to Des Bosses:

> accurately speaking, in place of "infinite number" we should say that more things are present than can be expressed by any number, or in place of "infinite straight line," that a line is extended beyond any specifiable magnitude, so that there already remains a longer and longer line. It is of the essence of number, of line, and of any whole whatsoever to be bounded. (Leibniz to Des Bosses, March 11, 1706; LDB 33)

Richard Arthur has argued that Leibniz developed this interpretation during his stay in Paris. As Arthur notes in his recent book,

> by the time he had finished laying the groundwork for his calculus in 1676, he already had an interpretation of the mathematical infinite and the infinitely small as *fictions*—as a means of abbreviating statements about an arbitrary large number of things of arbitrary smallness. Just as the infinite is not an actually existing whole made up of finite parts, so infinitesimals are not actually existing parts which can be composed into a finite whole. Borrowing a term from the Scholastics, Leibniz called the infinite and the infinitely small *syncategorematic terms*: like "it" and or "some" in a meaningful sentence, they do not in themselves refer to determinate things, but can be used perfectly meaningfully in a specified context. (Arthur 2014, 81)[20]

As Arthur (following Moore) glosses this elsewhere, a syncategorematic interpretation holds that, for any member in an infinite series (x), there is a member (y) larger than it; a categorematic interpretation holds that there is a member (y) which is larger

---

[20] Cf. Leibniz to Varignon, February 2, 1702; GM IV:93, L 543. Also, Leibniz to Bernoulli, June 7, 1698, GM III:499; L 511.

than any member (x) in an infinite series.[21] According to the categorematic position, there would be a largest number and a final term in an infinite series; according to the syncategorematic position, all numbers are finite, and "infinite" means larger (or smaller) than any assignable term.

As Leibniz writes to Bernoulli in 1699,

> Given infinitely many terms, it does not follow that there must be an infinitesimal term.... I concede the infinite multiplicity of terms, but this multiplicity does not constitute a number or a single whole. It signifies nothing but that there are more terms than can be designated by a number. Just so, there is a multiplicity or complex of numbers, but this multiplicity is not a number or a single whole.[22]

In the rest of this book, I refer to Leibniz's attitude toward the quantitative aspect of infinity (after 1676) as a syncategorematic approach, along the lines sketched here. There is, however, a controversy about the extent to which this interpretation applies to Leibniz's metaphysical context. Maria Rosa Antognazza has recently argued that "[t]his kind of infinite concerns the abstract, ideal entities treated by mathematics" (Antognazza 2015, 8).[23] Arthur and Rabouin hold that the syncategorematic interpretation applies more generally.[24] My view is that the syncategorematic interpretation applies primarily to the quantitative aspect of infinity; and thus it applies both to numbers and to substances or to any plurality or quantity or more precisely to anything that can be quantified. Its application is therefore broader than just the realm of possible, abstract, ideal entities treated by mathematics. But the application

---

[21] Arthur (2001, 107). Arthur further notes that "Infinitesimals are fictions in the sense that the terms designating them can be treated *as if* they refer to entities incomparably smaller than finite quantities, but really stand for variable finite quantities that can be taken as small as desired" (107). For more discussion regarding this reading, see O. B. Bassler, "Leibniz on the Indefinite as Infinite," *Review of Metaphysics* 51 (1998): 849–79; S. Levey, "Archimedes, Infinitesimals and the Law of Continuity: On Leibniz's Fictionalism," in *Infinitesimal Differences: Controversies Between Leibniz and His Contemporaries*, ed. U. Goldenbaum and D. Jesseph (Berlin and New York: Walter de Gruyter, 2008),107–34; and R. T. W. Arthur, "Leery Bedfellows: Newton and Leibniz on the Status of Infinitesimals," in *Infinitesimal Differences: Controversies Between Leibniz and His Contemporaries*, ed. U. Goldenbaum and D. Jesseph (Berlin and New York: Walter de Gruyter, 2008), 7–30.

[22] Leibniz to Bernoulli, February 21, 1699; GM III:575; L 514; translation from S. Levey, "Leibniz's Constructivism and Infinitely Folded Matter," in *New Essays on the Rationalists*, ed. R. Gennaro and C. Huenemann (New York: Oxford University Press, 1999), 139.

[23] See also Antognazza (2015, 9–10).

[24] R. T. W. Arthur and D. Rabouin, "Leibniz's Infinitesimals: A Clarification and Defence of the Syncategorematic Interpretation," forthcoming; see esp. sec. 3.

of the syncategorematic interpretation to the two other degrees of infinity noted here is less obvious. In particular, it does not seem sufficient to address some of the metaphysical implications of ascribing infinity both to God and to created substances, especially if these metaphysical contexts employ a qualitative sense of infinity. This employment seems rather clear in the context of God's infinity and somewhat less clear regarding the infinity of created substances. Both these contexts are developed throughout this book.

1.4 INFINITE NUMBER, THE LAW OF THE SERIES, AND FORCE

1.4.1 Infinite Number and Infinite Series

While Leibniz rejects the notion of an infinite number, he clearly accepts the notion of infinite series. The notion of a series plays a central role in the development of his mathematical work in general, and in the development of his infinitesimal calculus in particular. But what is the difference that Leibniz sees between infinite series and the notion of infinite number, such that the one is regarded as consistent while the other is regarded as inconsistent? One crucial difference is this: the notion of an infinite series avoids the contradiction of an infinite number because it is not defined as a collection of units; rather, it is defined through its law of generation or its rule of production. As Leibniz puts it, "[a] series is a multitude with a rule of order" (A 6.4:1426). Such a rule (*regula*) of order can be regarded as a kind of generator, whose activity produces ordered results. Leibniz employs here the same notion of generative definition that he uses to provide other real definitions—that is, by demonstrating that a given concept is consistent by showing how it is produced. In this sense, to define x is to show how the concept of x can be constructed, which also demonstrates its self-consistency. As we shall see in detail in chapter 2, this attitude is also exemplified in Leibniz's demand for proving the possibility of the "*Ens Perfectissimum*" and in disproving the possibility of an infinite number.

Here, for simplicity's sake, let us contrast two notions that on first sight seem to be very similar: the notion of an infinite number and the series of natural numbers. It is crucial to observe that Leibniz does not see the series of natural numbers as a collection of units; rather, he provides an operative definition. His definition is given by a procedure that *generates* the series, so that the successive number results from consecutive operations, viz., additions of units (or ones) (e.g., NE 4.2.1). The reiterability of this procedure implies that it can be carried on without limitation, but it does not imply that there is an infinitely large number. For Leibniz, this generative definition shows the intelligibility of the series of natural numbers.

This also reveals another feature of Leibniz's analysis of infinity in this context: namely, that it is bound up with a notion of activity or, more precisely, with the notion of the potential activity of an agent—namely, the possibility of continuing to carry out the procedure, without an upper bound. Thus, while an infinite series is not seen as a sum of units—that is, it is not a whole—one can speak of the law generating a series, seen as a definition *in act*, and at the same time as what gives the series its unity and intelligibility. Something of this attitude is expressed when Leibniz notes that "numbers do not *in themselves* go absolutely to infinity, since then there would be an infinite number" (A 6.3:503; LLC 99). An infinite series requires a rule whose activity would generate it. Thanks to its formation law, one can speak of a series, rather than seeing it as an aggregate or a collection of units (which would be the case for a finite number).

It is instructive to recall here an insightful remark by Yvon Belaval, who suggests that Leibniz's operative definition of number implies that

> [n]umber, as a relation, has no reality (and possibility) unless it is being thought. It is therefore on the side of the mind [*du côté de l'esprit*] that we have to look for the source of numerical infinity. It is the actual infinity of a mind [*l'esprit*] that accounts for the virtual infinity of any numerical series.[25]

Belaval's remark brings out another important feature of Leibniz's views on infinity; it suggests that numerical infinity is understood in the context of the actual thoughts of an active mind. In this sense, the reality of an infinite series, as well as other concepts involving infinity, presupposes the reality (and activity) of the mind thinking it. This sort of conceptualism about abstract entities, such as number and relations, is a familiar theme in Leibniz's thought from his early writings onward.[26] It also has far-reaching consequences for our current purposes of distinguishing between infinite beings and infinite nonbeings—that is, *entia rationis* such as infinite series. In light

---

[25] "[L]e nombre, tant que relation, n'a de réalité (et de possibilité) que pour autant qu'il est pensé. C'est donc du côté de l'esprit qu'il faut chercher la source de l'infinité numérique: dune manière plus précise, c'est l'infinité actuelle de l'esprit qui rend compte de l'infinité virtuelle de toute suite numérique" (Y. Belaval, *Leibniz critique de Descartes* [Paris: Gallimard, 1960], 270).

[26] As Couturat nicely points out, "one can say that Leibniz remains a nominalist in an entirely negative sense, namely that he rejects realism and denies universals a real and substantial existence. But he does not thereby refuse to assign them objective value, like the nominalists who reduce them to names. Rather, he adapts an intermediate position, which one designates by the name *conceptualism*" (L. Couturat, *La logique de Leibniz* [Hildesheim: Olms, 1961], 471; my translation). As Mugnai notes, "there are no ideas without the intellectual activity of someone thinking (be it God or man or some other rational being)" (M. Mugnai, "Leibniz's Theory of Relations," *Studia Leibnitiana* 28 [Suppl. 1992]: 25).

of this operative definition, the series of natural numbers can be seen as infinite in the context of presupposing a mind that could go on to add units that would make up larger and larger numbers. In other words, it looks as though Leibniz's real definition (through the method of production) also presupposes a producing (thinking) agent, which may be human or divine.[27] As we shall see, both infinity and agency (or activity) play an essential role in Leibniz's view of individual substance.

This conceptualist approach also informs Leibniz's position that infinity is to be regarded as merely potential in the contexts of abstract and mathematical concepts, and that this type of potential infinity does not apply to the infinity of true beings. True beings are agents whose activity is not merely potential but also actual. If their nature is to act, and if they never perish, then one might say that they always act, and they do so according to their intrinsic law—a law whose application generates the sequence of predicates that defines their very nature. In section 8.4, I will suggest that this is consistent both with the observation that the infinity of beings is not merely quantitative and with Leibniz's ascribing primitive force (or a source of life) to created beings.

1.4.2 The Law of the Series

Like the generative definition of the natural numbers, the notion of the "law of a series" is central both to Leibniz's metaphysics in general and to the role infinity plays in his view of substance in particular. As he writes apropos Foucher's comments in 1676, "[t]he essence of substances consist in the primitive force of acting (*la force primitive d'agir*), or in the law of the series of changes, like the nature of series in numbers" (A 6.3:326).[28] Leibniz is using the notion of the law of a series to define not only a series of numbers but also the very nature of true beings—that is, the unfolding states of individual substances.[29] Leibniz holds that every created being—that is, every

---

[27] As Leibniz writes, "An infinite extended thing is only imaginary. An infinite thinking thing is God himself" ("Comments on Spinoza's Philosophy," 1707?; AG 276).

[28] "L'essence des substances consiste dans la force primitive d'agir, ou dans la loy de la suite des changemens, comme la nature de la series dans les nombres" (A 6.3:326).

[29] "The law of order . . . constitutes the individuality of each particular substance" (GP IV:518; L 493). In a letter to De Volder in 1704, Leibniz writes, "[f]or me nothing is permanent in those things except the very law that involves the continued succession" (LDV 289), and "[t]he succeeding substance is held to be the same when the same law of the series, or of continuous simple transition, persists; which is what produces our belief that the subject of change, or monad, is the same. That there should be such a persistent law, which involves the future states of that which we conceive to be the same, is exactly what I say constitutes it as the same substance. . . . The fact that a certain law persists, which involves the future states of what we conceive to be the same – this is the very fact, I say, that constitutes that same substance" (GP II:263–64; L 534–35; LDV 291). See also A 6.4:1673; and *Monadologie* 22. See also J. Jorati, *Leibniz on Causation and Agency* (Cambridge: Cambridge University Press, 2017), 4–16.

individual substance—is defined through its own law of production, which he likens to a law of the series.[30] The law accounts for the individuality of a created substance, as well as for its identity over time.[31] This indicates that, for Leibniz, all beings other than God are infinite in the sense that they are defined and individuated through an infinite (and unique) law of generation. The term "generation" has a dual sense in this context: it applies both to the generation of the individual's concept (in God's mind) and to its realization in the world once it is created and given power to act.

We know that, in his middle years, Leibniz developed the view that each individual has a complete concept, a concept that entails all its predicates. As Leibniz argues in the "Discourse on Metaphysics," articles 8 and 13, the complete concept of the individual includes everything true of it. Such a concept is clearly infinite as well. The comparison between an individual's complete concept and the individual shows another sharp contrast between an infinite being (that is, an individual substance) and an infinite nonbeing, viz., its conceptual counterpart (that is, its complete concept). In effect, this is a clear articulation of the contrast between concepts and agents in Leibniz's metaphysics. It goes without saying that a complete concept of an individual must be consistent. If an individual exists, as many surely do, then it must also be possible—that is, it must have a consistent concept or a perfect representation in God's understanding. Indeed, in the "Discourse on Metaphysics" and the ensuing correspondence with Arnauld, Leibniz illustrates such cases with examples of historical individuals, such as Alexander the Great and Julius Caesar.[32]

Leibniz, however, is not so explicit about how such infinite concepts of individuals are to be defined. In previous work,[33] I have argued that Leibniz sees complete concepts in terms of their real definitions—that is, through their method of production or the way they are thought in God's understanding. Leibniz is very explicit that

---

[30] In the Theodicy §291, Leibniz writes, "[in my *system of pre-established harmony*,] I pointed out that by nature every simple substance has perception, and that its individuality consists in the perpetual law which brings about the sequence of perceptions that are assigned to it, springing naturally from one another" (Theodicy, p. 304).

[31] "[T]here must necessarily be a reason allowing us truly to say that we endure, that is to say, that I, who was in Paris, am now in Germany. For if there is no such reason, we would have as much right to say that it is someone else. . . . [T]hese predicates were laws included in the subject or in my complete notion, which constitutes what is called I, which is the foundation of the connection of all my different states and which God has known perfectly from all eternity" (remarks on Arnauld's letter, May 1686; AG 73; GP II:43).

[32] A curious question that Fichant raises is why these references to such individuals are not present in the texts of the monadological period. See M. Fichant, "L'invention métaphysique," introduction to *G. W. Leibniz: Discours de métaphisuque uivi de Monadologie et autres textes*, ed. M. Fichant (Paris: Gallimard, 2004). I attend to this question in section 8.3, this volume.

[33] O. Nachtomy, *Possibility, Agency, and Individuality in Leibniz's Metaphysics*, New Synthese Historical Library (Dordrecht: Springer, 2007a), ch. 2.

the concept of an individual involves infinitely many predicates, which all serve to define the whole career of an individual—past, present, and future. Thus, each predicate is an essential component of the definition of that individual.[34] However, if such a concept would be seen as a mere conjunction of predicates or as a set of infinitely many predicates, it would be contradictory, just as any determinate infinite number of units is. For this reason, it seems that Leibniz thought about the infinite concept of an individual along the lines that he thought about an infinite series—that is, as defined by its law of generation.[35] In fact, I take it that this is why Leibniz found the notion of the law of the series so attractive in the first place and why it plays such a central role in his metaphysics.[36] What I would like to stress here is that, in either version, the notion of a law or rule of production enables Leibniz to account for the infinity of individual substance without falling prey to the inconsistency involved in the notion of infinite number.

1.4.3 The Law of the Series, Primitive Force, and Derivative Force

In a letter to De Volder, Leibniz notes,

> [a]ll individual things are successive, i.e., are subject to succession. . . . For me, nothing is permanent in those things except the very law that involves the continued succession, which in individual things corresponds to the law that is in the whole universe. (January 21, 1704; LDV 289)[37]

---

[34] See Fichant (2004, 51–52), for an explication of this point.

[35] See Nachtomy (2007a, chs. 1 and 2).

[36] The story of this concept within Leibniz's development, however, is rather complex. Fichant has argued that, in the transition from the "Discourse on Metaphysics" to the *Monadologie*, the law of the series takes the role of the complete concept of the individual. As Fichant writes, "in the new theory of substance developed towards the *Monadology*, the concept of *a series of operations* replaces that of *a complete notion entailing all predicates*" (2004, 528n10). Fichant observes that, in the late writings, the terminology of individual substance and the *in esse* principle, along with the historical examples of the "Discourse on Metaphysics" disappear (2004, 114–15). From a textual point of view, this observation is certainly correct. At the same time, it is worth pointing out that a notion of a production rule or a method of production is operative even before the "Discourse on Metaphysics," in the more basic sense of producing the very complete concept of an individual as a possibility in God's mind. And this concept surely remains operative in Leibniz's later texts as well.

[37] "The substance that succeeds is taken to be the same as long as the same law of the series, i.e., of the continual simple transition, persists that gives rise to our belief in the same subject of change, i.e., the monad. I say that the fact that there is a certain persisting law, which involves the future states of that which we conceive as the same, is the very thing that constitutes the same substance" (Letter to De Volder, January 21, 1704; LDV 291).

In his correspondence with De Volder, it is clear that Leibniz sees a close relation between the law of the series and the primitive and derivative forces that also serve to define individual substances:

> derivative force is the present state itself in so far as it tends toward a following state, i.e., pre-involves a following state, just as everything in the present is pregnant with the future. But the persisting thing itself, insofar as it involves all cases, has primitive force, so that primitive force is like the law of a series, and derivative force is like a determination that designates some term in the series. (LDV 287; GP II:262)

"And indeed, derivative forces are nothing but modifications and echoes of primitive forces" (LDV 263).[38] In particular, primitive force is likened to the law of the series, and derivative force to a determination that designates some term in the series. Roughly speaking, primitive force corresponds to the principle of change (which persists) and derivative force corresponds to the actual state of the substance (designated by the current term in the series); primitive force relates to the law that generates change, and derivative force to the current, changing state. In his later texts, Leibniz articulates this relation in terms of appetition and perception, so that appetite is the drive from one perception to another. Appetite, like primitive force, seems to be constant and continuous, while the various states (or perceptions) are discrete and ever changing. Likewise, the law is one and its states are varying and proceed to infinity.[39]

The way Leibniz conjoins primitive force and the law of the series is of great interest. But, unfortunately, he is not entirely clear about it. In my view, he is conjoining two aspects, such that the *law* provides a course of action and *primitive force* provides power of action. The law provides the information as to what to do, and the force provides the power that enables its execution. Each of these two components is a necessary condition for Leibniz's view of individual agents. In itself, the law can be seen as pointing to a possible individual (conceived in God's mind) and the law *along with* primitive force can be seen as pointing to an actual individual. Indeed, as Leibniz writes,

---

[38] For further illuminating discussion of the relation between primitive and derivative forces, see R. M. Adams, *Leibniz: Determinist, Theist, Idealist* (New York: Oxford University Press, 1994), 379–80; D. Garber, "Leibniz, Physics and Philosophy," in *The Cambridge Companion to Leibniz*, ed. N. Jolley (Cambridge: Cambridge University Press, 1995), 292–93.

[39] As Phemister puts this, "[p]rimitive force is, as it were, the blueprint of the living animal. It is the law that contains the information required to generate the creature's entire life experiences. Mentally, these experiences are perceptions; physically, they are derivative forces that together propel the organic body into motion" (P. Phemister, *Leibniz and the Natural World Activity, Passivity and Corporeal Substances in Leibniz's Philosophy* [Dordrecht: Springer, 2005], 195).

> For to say that, in creation, God gave bodies a law for acting means nothing, unless, at the same time, he gave them something by means of which it could happen that the law is followed, otherwise, he himself would always have to look after carrying out the law in an extraordinary way. But indeed, this law is efficacious, and he did render bodies efficacious, that is, he gave them an inherent force. (1702; AG 253–54; GP IV:395–96)

Created beings are efficacious owing to their inherent force; they are agents endowed with a divine-like source of action—primitive active force—that does not diminish as long as they act. Once created, substances are active owing to an inherent source of action—that is, primitive force or life;[40] they always act, but their source of action does not decrease. In this sense, primitive force is not seen as a quantity; for otherwise, it would be finite and would diminish. But since it does not change—it neither decreases nor increases—it may be seen as an infinite source of action; it would certainly go on as long as a substance lives, which is forever. At the same time, the force God placed in created things is obviously not the same as God's absolute power; rather, while infinite, the primitive active force ought to be regarded as infinite (or omnipotent) in a lower degree. In chapter 8, I develop the suggestion that Leibniz's notion of primitive force (as conjoined with the law of the series) supports the view of created substances as infinite in kind.

---

[40] I develop the analogy between primitive force and life in chapter 10, this volume.

2

LEIBNIZ IN PARIS
BETWEEN INFINITE NUMBER AND
INFINITE BEING

TH: So doesn't even God understand the number of all unities?
PA: How do you suppose he understands what is impossible?
—G. W. Leibniz[1]

This chapter presents two major steps in the development of Leibniz's views on infinity through his responses to Galileo and Descartes. Section 2.1 presents Leibniz's response to Galileo's paradox—that is, his rejection of infinite number. Section 2.2 presents a problem that arises from Leibniz's resolution of Galileo's paradox. Briefly, the problem is this: If Leibniz regards the notion of infinite number as inconsistent, how is it that he regards the notion of infinite being as consistent? In section 2.3, I consider a semantic solution to this problem and conclude that it is appealing but ultimately inadequate. In section 2.4, I consider a more promising solution—namely, that Leibniz distinguishes between different senses of infinity. The chapter concludes (section 2.5) with a discussion of Leibniz's attitude toward infinity vis-à-vis his critique of Descartes's distinction between the infinite and the indefinite.

2.1 GALILEO'S PARADOX AND LEIBNIZ'S RESPONSE

The following note on Galileo's *Two New Sciences* encapsulates Leibniz's response to the paradoxes presented by Galileo:

> Among numbers there are infinite roots, infinite squares, infinite cubes. Moreover, there are as many square numbers as there are numbers in the universe. Which is impossible. Hence it follows either that in the infinite

---

[1] "Pacidius to Philalethes: A First Philosophy of Motion," October 29–November 10, 1676; A 6.3:552; LLC 181.

the whole is not greater than the part, which is the opinion of Galileo and Gregory of St. Vincent, and which I cannot accept; or that infinity itself is nothing, i.e. that it is not one and not a whole. (Fall 1672; A 6.3:168; LLC 9)

In his *Discourses and Mathematical Demonstrations Concerning the Two New Sciences*, Galileo presents several paradoxes concerning the infinite. He opens the discussion with a geometrical example of the turning wheel (sometimes referred to as *rota Aristotelis*; LLC 432). Galileo is interested in comparing the lengths of the lines drawn by the perimeter as the wheel is turning with that drawn by the point at the wheel's center (or, indeed, with any point in the wheel; EN 68). It turns out that the lengths of the lines drawn by any point on the wheel's radius as it revolves are equal. Since even the line drawn by the center of the circle is equal to a line drawn on a radius length at infinity, the example illustrates the paradoxical result that the smallest number is equal to an infinite one (as they all yield lines of equal length).

Galileo then provides another arithmetical argument to make a similar point. This argument, which Leibniz cites in several places,[2] is referred to in the recent literature as "Galileo's paradox." It runs as follows: "all numbers, comprising the squares and the non-squares, are greater than the squares alone" (EN 78; LLC 356). In other words, the series 1, 2, 3, 4, 5, 6, ... (of squares and non-squares) has members that the series of squares alone 1, 4, 9, 16, ... does not have (2, 3, 5, 6,). But, "there are as many square numbers as there are their own roots, since every square has its own root, and every root its own square.... But, if I were to ask how many roots there are, it cannot be denied that there are as many as all the numbers, because there is no number that is not a root of some square. That being so, it must be said that the square numbers are as many as all the numbers, because they are as many as their roots, and all numbers are roots" (EN 78; LLC 356). On the one hand, there appear to be more numbers than squares, but on the other hand, there are as many numbers as squares. Thus, it turns out that the quantity of squares is both "less than" and "equal to" the quantity of all numbers—a paradox. Given this paradox, one might be inclined to infer that the relations of "greater than," "less than," and "equal to" do not apply in the context of infinity. This was Galileo's conclusion.

The radical conclusion Galileo draws from this argument is reiterated by Salviati: "from your ingenious argument we are led to conclude that the attributes 'larger,' 'smaller,' and 'equal' have no place either in comparing infinite quantities with each other or in comparing infinite with finite quantities" (EN 80). As Galileo

---

[2] E.g., "Pacidius to Philalethes"; LLC 179.

writes, "I believe that these attributes of greatness, smallness, and equality do not befit infinities, about which it cannot be said that one is greater than, smaller than, or equal to one another" (EN 77–78; LLC 355).

Galileo concludes that insurmountable paradoxes arise when the notion of infinity is regarded as a quantity. The paradoxes he points to show that the most basic properties that must pertain to a quantity (such as "bigger than," "smaller than," or "equal to") do not hold in the case of "infinite quantity" (EN 80). As Knobloch has stressed, if this is the case, infinities should not be regarded as *quanta* at all.[3]

Leibniz's conclusion from his readings of Galileo's *Two New Sciences* in 1672–73, however, was rather different. Instead of determining that the infinite does not belong in the realm of quantity, Leibniz comes to the conclusion that the notion of *an infinite number, seen as a whole*, is impossible. It is impossible precisely because such a notion violates the axioms of the realm of quantity—more specifically, the axiom stating that the whole is greater than its part.[4]

Leibniz argues that one cannot accept the result that the series of natural numbers is equal to the series of their squares; for if this were permitted, the whole (the series of natural numbers) would not be greater than its parts (the series of squares). However, he finds "it difficult to agree" with Galileo's conclusion that the "appellations of greater, equal, and less have no place in the infinite" (A 6.3:551; LLC 179), "[f]or who would deny that number of square numbers are contained in the number of all numbers? But to be contained in something is certainly to be a part of it, and I believe it to be no less true in the infinite than in the finite that the part is less than the whole" (A 6.3:551; LLC 179).

Richard Arthur nicely presents the choices Leibniz sees as emerging from Galileo's paradox in the form of the following dilemma:

> [Leibniz] identifies two candidates for rejection: (W) that in the infinite the whole is greater than the part, and (C) that an infinite collection (such as the set of all numbers) is a whole or unity. . . . Leibniz upholds W, and this leads him to reject C. Cantor upholds C, and this leads him to reject W. (Arthur 2001, 103–104)

---

[3] E. Knobloch, "Galileo and Leibniz: Different Approaches to Infinity," *Archive for the History of the Exact Sciences* 54 (1999): 87–99.

[4] "[J]ust as the proposition 'the whole is greater than the part' is the basis of arithmetic and geometry, i.e., of the sciences of quantity, similarly, the proposition 'nothing exists without reason' is the foundation of physics and morality, i.e., the sciences of quality, or, what is the same (for quality is nothing but the power of acting and being acted on) the sciences of action, including thought and action" (*Confessio* 35; A 6.3:118). See also Knobloch (1999, 94).

Leibniz concludes that the whole is greater than the parts even for the infinite, and therefore must deny that an infinite number is a whole. Thus, according to Leibniz, there cannot be a number of all numbers, or an infinite number. This implies that an infinite collection of elements cannot be regarded as a genuine unity.

As noted, the conclusion Galileo draws from his paradoxes is that, if the most basic relations of quantity do not hold in the realm of infinity, then infinity cannot be regarded as a quantity. This further implies that the finite and the infinite belong to different categories that cannot even be compared. Knobloch (1999) has put this point as follows. He argues that, according to Galileo, "[A]n 'infinite quantity' would ... be a 'contradiction in terms,' because an infinite would lack precisely those properties that characterize a quantity" (94).

Knobloch maintains that Leibniz's response to Galileo was to show, through his mathematical work on the calculus, that infinities can be handled in quantitative and precise terms. However, as I will argue, while Knobloch's account is correct, it leaves out an important part of the story. Knobloch observes that Leibniz's calculus and his interpretation thereof show how the infinite can be dealt with mathematically—that is, as a quantity. This is no doubt true. However, this observation holds only for *one* sense of infinity, which Leibniz (not surprisingly) reserves for quantities, numbers, and magnitudes (and for which he develops his syncategorematic approach). But there is another sense of infinity for which Leibniz actually accepts Galileo's position that the infinite cannot be regarded as a quantity. This is the notion of infinity that he would apply to being in general, and to God's being (and perfection) in particular. I will argue that this distinction, between a quantitative and a nonquantitative sense of infinity, is of great consequence for Leibniz's resolution of the paradoxes of infinity and for his wider metaphysics.

To a large extent, Leibniz's approach to infinity can be seen as a complex response to, and a sophisticated development of, Galileo's conclusions. In working his way through Galileo's *Dialogues*, Leibniz has "acquired" two challenges: (1) On the one hand, his recognition of Galileo's paradox motivates him to distinguish between a kind of infinity that he regards as nonquantitative and applicable to beings (and especially to the most perfect being) and a kind of infinity that is quantitative and applicable to the mathematical domain. (2) This immediately sets another challenge for Leibniz: to show how one can treat the notion of infinity within mathematics—that is, in a quantitative sense. Much of Leibniz's work on infinite series and the calculus during his Paris years can be seen as a response to this task. One result of his efforts is the syncategorematic interpretation of infinite terms (seen as fictions) that we glossed in the previous chapter.

At the same time, Leibniz's resolution of Galileo's paradox in terms of rejecting infinite number gives rise to another problem. That problem is how to account for the difference between the notion of an infinite number (which he regards as impossible) and the notion of an infinite being (the primary and most perfect being), which he regards not only as possible but also as implying a necessary being—one whose nonexistence is impossible. How to account for the difference between these two notions is what I call "Leibniz's problem."

2.2 LEIBNIZ'S PROBLEM: INFINITE BEING AND INFINITE NUMBER

Leibniz's claim that "the number of all numbers is a contradiction" (e.g., A 6.3:463; DSR 7) appears in his Paris notes from 1675–76, a period during which he was developing his views about infinity in various domains. At the same time, Leibniz was also engaged, among many other projects, in distinguishing between possible and impossible notions. It is well known that Leibniz's view of possibility plays a central role in his metaphysics.[5] As early as his "Confession of a Philosopher" (1672–73), Leibniz defines a possibility such that x is possible if it has a notion whose internal constituents are consistent. In this context, Leibniz is using the notions of the "number of all numbers" (*numerum omnium numerorum*) and the "greatest or maximal number" (*numerus maximus*) as an illustration of an impossible notion—that is, a notion whose internal constituents imply a contradiction. In the same texts, Leibniz also uses the notion of the number of all numbers in contrast to a notion whose possibility he is keen to prove—that of "the greatest or the most perfect being" (A 6.3:572; DSR 91).

Comparing the notion of the greatest being with the notion of the greatest number gives rise to a severe problem. Leibniz states this problem in a letter to Oldenburg from December 1675:

> Whatever the conclusions which the Scholastics . . . and others derived from the concept of that being whose essence is to exist, they remain weak as long as it is not established whether such being is possible, provided it can be thought. To assert such a thing is easy; to understand it is not so easy. Assuming that such a being is possible or that there is some idea corresponding to these words, it certainly follows that such a being exists. But we believe that we are thinking of many things (though confusedly) which

[5] For more details regarding Leibniz's view of possibility, see Nachtomy (2007a).

nevertheless imply a contradiction; for example, the number of all numbers. We ought strongly to suspect the concepts of infinity, of maximum and minimum, of the most perfect, and of allness (*omnia*) itself. Nor ought we believe in such concepts until they have been tested by that criterion which must, I believe, be credited to me, and which renders truth stable, visible and irresistible. (GM I:83–84; L 257)

The worry raised by Leibniz here is made even clearer in a letter to Countess Elizabeth, written three years later (1678). There, Leibniz considers several examples of impossible notions (such as those of the squared circle and of the greatest speed) and writes,

we think about this greatest speed, something that has no idea since it is impossible. Similarly, the greatest circle of all is an impossible thing, and the number of all possible units is no less so; we have a demonstration of this. And nevertheless, we think about all this. That is why there are surely grounds for wondering whether we should be careful about the idea of the greatest of all beings, and whether it might not contain a contradiction. (A 2.1:433–38; AG 238)

Leibniz's reasoning here is very clear. Since we entertain thoughts about things such as the greatest speed and the greatest number, which upon analysis prove to be contradictory, we ought to examine whether the idea of the greatest of all beings might not turn out to be contradictory as well. In fact, we often use concatenations of words that do not correspond to any idea and which might well turn out to be contradictory.

Leibniz's problem is therefore to show that, while the greatest number is contradictory and thus impossible, the greatest or most perfect Being—that is, God—is not. In fact, Leibniz's response to this problem turns out to encompass two distinct phases: first, showing that *the concept of God* is not contradictory and is therefore possible; and second, showing that if God's existence is possible, it is also necessary.

The way in which Leibniz compares and contrasts the notions of infinite number and that of the infinite (or most perfect) being has drawn surprisingly little attention from scholars. Even more important, the significance of this comparison for Leibniz's views about infinity has gone almost unnoticed. In this and the following chapter, I will present Leibniz's approach to infinity through his response to this problem. It is not easy to see how Leibniz convinced himself that, while the notion of the greatest number is self-contradictory, the notion of the greatest being is not. Formulating this dilemma in detail will set the stage for exploring Leibniz's solution to this problem through his encounter with Spinoza in the following chapter.

2.2.1 The Context

As noted, Leibniz's treatment of Galileo's paradox occurs in 1675–76—shortly after the period in which he develops his calculus and while he is working out his interpretation of these results (especially regarding the status of infinitesimals). In the very same period, Leibniz is also engaged in a project of distinguishing possible and impossible notions. In fact, Leibniz explicitly attempts to *demonstrate* that some notions are possible while others are not. In particular, he argues that, in order to show that the most perfect being exists, one has to show first that its notion is possible (i.e., that it has a consistent notion). This demonstrates the important role that possibility proofs—which he also calls "causal" or "genetic" definitions—play in Leibniz's early thought.

By that time, Leibniz was already working with a fairly well-crystallized set of presuppositions regarding the notion of possibility:[6] he identifies the possible with that which is conceivable in God's mind, and he explicates the conceivable (or the intelligible) in terms of the consistency relations between the terms that make complex notions.[7] Thus, for example, in the same set of notes from this period, Leibniz states that "every thing possible is thinkable" (A 6.3:475; DSR 29).[8] He was also working with a general method for distinguishing between possible and impossible notions. His method can be stated roughly as follows: if the terms whose composition makes up a given notion are consistent *inter se*, then this notion indicates a genuine possibility.[9] But if the terms whose would-be composition makes up a given notion are inconsistent, such that they imply a contradiction, then the notion indicates an impossibility or an impossible thing—that is, something that cannot exist. In this case, there is no concept corresponding to it in God's mind but only a concatenation of words or signs in human minds, to which no notion corresponds in God's mind. The

---

[6] I have written extensively on this issue in my (2007a) book and elsewhere, and so here I will just make some very brief remarks.

[7] "I have defined the necessary as that the contrary of which cannot be understood; therefore, necessity and impossibility of things are to be sought in the ideas of things themselves, and not outside those things, by examining whether they can be thought or whether they imply a contradiction" (*Confessio* 57; G. W. Leibniz, *Confessio philosophi*, ed. and trans. Y. Belaval [Paris: J. Vrin, 1970], 56).

[8] "Those things are contingent that are not necessary; those are possible whose non-existence is not necessary. Those are impossible that are not possible, or more briefly: the possible is what can be conceived, that is (in order that the word "can" does not occur in the definition of possible) what is understood clearly by an attentive mind; the impossible—what is not possible" (*Confessio* 55; Leibniz 1970, 54).

[9] Strictly speaking, this condition only applies to what is logically possible or possible per se. The notion of compossibility is more restrictive: it depends on the possibility of coexisting with other individuals in the same world.

method of showing whether a certain concatenation of terms indicates a real possibility or not requires an analysis of complex notions into their simple constituents in order to determine whether they involve internal contradictions or not.

In his Paris notes, Leibniz frequently uses the notion of the number of all numbers to illustrate an impossible notion. For example, he states that "[t]he number of all numbers is a contradiction, i.e., there is no idea of it; for otherwise it would follow that the whole is equal to the part, or that there are as many numbers as there are square numbers" (A 6.3:463; DSR 7). He then immediately moves to discuss the "twofold origin of impossibility":

> Impossible is a twofold concept [*duplex notio*]: that which does not have essence, and that which does not have existence, i.e., that which neither was, is, nor will be because it is incompatible with God, or, with the existence or reason which brings it about that things exist rather than do not exist. One must see if it can be proved that there are essences which lack existence. . . . The origin of impossibility is twofold: one from essence, the other from existence or, positing as actual. (A 6.3:464; DSR 7)

The "number of all numbers" clearly belongs to the first type of impossibility. If it is contradictory, it is logically impossible and thus has no essence. Incidentally, note that Leibniz's strategy of defining essences in terms of consistent and intelligible notions is already at work here. For other notions may be logically possible—and so have an essence—while lacking existence, since they may be incompatible with the reason that "brings it about that things exist rather than do not exist."[10]

I have already noted that Leibniz's use of the notions of the greatest number, the greatest line, and the most rapid motion, all of which occur as examples of impossible notions, appear in a rather specific context—namely, in contrast to a notion whose possibility Leibniz is very keen to prove in the same texts, viz., the notion of the most perfect Being (the *Ens Pefectissimum*; A 6.3:572; DSR 91; A 6.3:325). Leibniz's immediate objective here is to support Anselm's argument, as revived by Descartes, according to which God's existence can be proved a priori by an analysis of its notion. The main assumption of this argument is that "existence" is included in the notion of the most perfect Being as one of its perfections. Leibniz however, argues that, for Descartes's argument to be valid, one first has to show that the very definition of God

---

[10] In the *Confessio*, Leibniz gives the following example for something that has essence but lacks existence: "a species of animal with an uneven number of feet" (A 6.3:128; *Confessio* 57). What Leibniz calls here impossibility from existence is very closely related to his notion of compossibility or the possible coexistence of different individuals.

as the most perfect Being is consistent. As he writes, "God is a being from whose possibility (or, from whose essence) his existence follows. If a God defined in this way is possible, it follows that he exists" (A 6.3:582; DSR 105).

Leibniz's contribution here is to question the modal status of the antecedent, thus giving the argument an explicit *modus ponens* form. In this way, Leibniz will attempt to prove the antecedent, so that the conclusion, "God exists," would readily follow. Note also that in lieu of the traditional appeal to the definition of God through its essence, Leibniz appeals to its possibility. This reflects a more general tenet of Leibniz's thought, according to which the essence of a thing can be expressed in terms of its possibility. In Leibniz's eyes, Descartes, as well as the tradition preceding him, have assumed (without proof) that the definition of the most perfect being is nonproblematic. Leibniz points out that this supposition is not at all self-evident, and thus requires a proof.

Now, as we have seen, according to Leibniz, definitions of concepts must pass the test of self-consistency or intelligibility—that is, their constituent elements must be shown to be free of internal contradictions. This is particularly clear in his critique of Anselm's demonstration of the existence of God. More generally, Leibniz's approach indicates that, in order for the truth or falsity of existence claims to be ascertained, their possibility or intelligibility has to be established first.[11] Establishing the possibility of a concept requires what Leibniz calls a "real definition," which he spells out as follows:[12]

> A real definition is one according to which it is established that the defined thing is possible, and does not imply a contradiction. For if this is not established for a given thing, then no reasoning can be safely taken about it, since if it involves a contradiction, the opposite can perhaps be concluded about the same thing with equal right. And this was the defect in Anselm's demonstration, revived by Descartes, that the most perfect or the greatest being must exist, since it involves existence. For it is assumed without proof that a most perfect being does not imply a contradiction; and this gave me occasion to recognize what the nature of real definition was. ("A Specimen of Discoveries," circa 1686; LLC 305–307)[13]

---

[11] In passing, let me note that the articulation of this point is one of Leibniz's significant contributions that has not received due attention.

[12] In 1684, Leibniz makes the connection between a real definition and a causal definition explicit. A real definition establishes the possibility of a concept; a causal definition reproduces the way it is produced in God's mind, and hence provides a much more thorough knowledge of it.

[13] This point is also supported in Leibniz's 1684 "Meditation on Knowledge, Truth and Ideas" (A 6.4:588–89; AG 25–26).

As Leibniz indicates here, his early critique of the notion of the most perfect being has led him to recognize what the nature of real definition is. Leibniz's critique of Descartes's use of the notion of the most perfect being is well known. But the fact that he was preoccupied with the *contrast* between the possibility of the most perfect being and the impossibility of the greatest number has slipped under the radar of most commentators. In the next section, I further develop this point, showing some more evidence that Leibniz was preoccupied with the comparison between the notion of the infinite (or most perfect) being and that of infinite number.

2.2.2 The Most Perfect Being and Infinite Number

Leibniz defines the notion of the most perfect being as "a subject of all perfections" (A 6.3:580; DSR 103). He further describes the subject of all perfections as "one which contains all essence, or which has all qualities, or all affirmative attributes" (A 6.3:572; DSR 91). In other words, the most perfect being is understood as the subject of all perfections or all positive attributes. Leibniz provides this definition in order to demonstrate that such a definition "is possible or [*seu*] does not imply a contradiction" (A 6.3:572; DSR 91). His strategy in providing such a possibility proof is to show that all positive and simple attributes (also referred to as perfections or forms) are compatible between themselves, such that they can all be attributes *of* the same subject.[14] As he writes, "[i]t will therefore be sufficient to have shown the compatibility of all primary or unanalyzable attributes" (A 6.3:572; DSR 91). If it can be shown that all perfections are compatible, it would follow that the notion of a most perfect being, seen as the subject of all positive attributes, is possible or intelligible, which is what he seeks to show at this point. Recall that this is the first step of the argument, in which what needs to be shown is that a most perfect Being is possible (*Quod Ens Perfectissimum sit possibile*, A 6.3:572). Once this is shown, the next step would be to show that such a being exists (*Quod Ens Perfectissimum existit*, A 6.3:578).

In these formulations, it is quite natural to read the notion of the most perfect being as analogous to a notion of an *infinite totality* of perfections or attributes. By slightly modifying traditional formulations, Leibniz defines God as that which contains all essences, all perfections, all qualities, or all affirmative attributes (e.g., in A 6.3:395–96, 572, 576).[15] In other words, in the definition of God as "the subject of *all* perfections or attributes," it is natural to read the "all" as indicating an infinite number of perfections.

---

[14] We can in fact detect here an early supposition of the *in esse* principle.
[15] This of course is not new. Descartes and Spinoza, for example, use formulations that are not too different.

This would be a rather plausible explication of the traditional definition of God as "that than which a greater cannot be conceived" and as an infinite being, which is explicit in both Descartes and Spinoza, among many others.[16] This impression is supported by considering Leibniz's argument that the notion of the *Ens Perfectissimum* is possible. As he writes,

> I seem to have discovered a demonstration that a most perfect being—or one which contains all essence, or which has all qualities, or all affirmative attributes—is possible, or does not imply a contradiction. This will be evident if I show that all (positive) attributes are compatible with each other. But attributes are either analyzable or unanalyzable; if they are analyzable they will be aggregates of those into which they are analyzed. It will therefore be sufficient to have shown the compatibility of all primary or unanalyzable attributes, or, of those which are conceived through themselves. For if individual attributes are compatible, so are several attributes, and so therefore are composite attributes. It will therefore be sufficient to show only the intelligibility of a being which contains all primary attributes, or, to show that any two primary attributes are compatible with each other. . . . Let these attributes be A and B. If they are incompatible, then the proposition "Quality A and quality B cannot be in the same subject" will be necessary, and so will be either an identical proposition or one which is demonstrable. It cannot be an identical proposition; for then "Where A is, B cannot be" would the same as "A is A" or "A is B," and so the one would express the exclusion of the other, and so one of them would be the negative of the other. But this is contrary to the hypothesis, for we have assumed that all attributes are affirmative. It is not demonstrable; for if one were to demonstrate, "Where A is, B cannot be," that could only be done by the analysis of one or other term, or of both. This is contrary to the hypothesis; for we have assumed them to be unanalyzable. Therefore, this incompatibility cannot be demonstrated. Therefore there is no incompatibility, and so any two affirmative qualities are compatible, and so all such qualities are compatible with all. Therefore a being which has all attributes is possible. (November? 1676; A 6.3:572; DSR 91–93)[17]

---

[16] Leibniz writes of Spinoza, "He defines God as an absolutely infinite being, likewise as a being that contains all perfections, i.e., affirmations, or realities or things that can be conceived" (A 6.3:384; LLC 43). However, in his letter 12, Spinoza makes clear that his characterization of the Substance as infinite does not pertain to the number of perfections. This is discussed in the next chapter.

[17] "Demonstrationem reperisse videor, quod Ens perfectissimum, seu quod omnem Essentiam contineat, seu quod omnes habeat Qualitates, seu omnia attributa affirmativa, sit possibile, seu non implicet contradictionem. Hoc patebit si ostendero omnia attributa (positiva) esse inter

While Leibniz's attempt to demonstrate the possibility of a most perfect being is certainly original, his characterization of the notion of a most perfect being is not. Rather, his definition is grounded in a familiar and traditional definition—one that goes back at least to Anselm. If Leibniz is working with a traditional and widely accepted definition of God, why is he worried about the intelligibility of a being that contains all primary attributes? Why should the possibility of the most perfect being—a traditionally accepted and apparently innocuous notion—require a proof? Why, in other words, should its possibility be in question?

A general answer to this question can be gathered from what we have already covered—that is, that Leibniz was engaged at the time in a project, related to the construction of his philosophical/rational language, of distinguishing between possible and impossible notions by means of the analysis of complex concepts. It seems, however, that Leibniz's particular interest in providing a possibility proof for the notion of the most perfect being ought to be seen against a more specific background. I suggest the following: Leibniz is concerned with the possibility of the notion of totality in general, and with the definition of God as the maximal totality in particular, because he observes that notions such as "infinite number" and "the most rapid motion" and "the greatest shape"—all of which have an obvious similarity with the notion of the greatest being—involve a contradiction.[18] The similarity between the notion of the greatest number, seen as the totality of all numbers, and that of the greatest being, seen as the totality of (infinitely many) perfections, is quite evident. Furthermore, Leibniz's argument presupposes a discrete notion of attributes, which may be enumerated as the units making up a number. This strongly supports an analogy between these notions. One can even think of showing this analogy by drawing a one-to-one correspondence between each of the units making up a number and the perfections included in the subject of the most perfect being.

---

se compatibilia. Sunt autem attributa aut resolubilia, aut irresolubilia, si resolubilia sunt erunt aggregatum eorum in quae resolvuntur; suffecerit ergo ostendisse compatibilitatem omnium primorum, sive irresolubilium attributorum, sive quae per se concipiuntur, ita enim si singula compatibilia erunt, etiam plura erunt, adeoque et composita. Tantum ergo suffecerit ostendere Ens intelligi posse, quod omnia attributa prima contineat, seu duo quaelibet attributa prima esse inter se compatibilia" (A 6.3:572; DSR 91–93).

[18] "There cannot be a most rapid motion or a greatest number. For number is something discrete, where the whole is not prior to its parts, but conversely. There cannot be a most rapid motion, because motion is a modification, and is the transference of a certain thing in a certain time. (Just as there cannot be a greatest shape.) There cannot be one motion of the whole, but there can be a kind of thinking of all things. Whenever the whole is prior to its parts, then it is a maximum, as in space and in a continuum. If matter is like a shape, namely that which makes a modification, then it seems that there is no totality of matter" (A 6.3:520; DSR 79).

Leibniz analyzes the notion of the greatest being in terms that indicate a maximal totality. For instance, he uses phrases such as "the subject of all perfections" (A 6.3:580; DSR 103) and "one which contains all essence, or which has all qualities, or all affirmative attributes." In a set of definitions from 1676, which relates to Euclid's definition of number, Leibniz writes, "[n]umber, if it is understood simply as integral and rational, is a whole consisting of units (*totum ex unitatibus constans*)" (A 6.3:482–84; DSR 37–39). In the same set of texts, he also draws an explicit analogy between God's essence and whole numbers. Accordingly, God's essence consists of simple forms or perfections as numbers consist of units.[19] Since Leibniz defines a whole number as one consisting of units, the greatest number (which he deems impossible) is seen as one that would consist of all units. Analogously, since he defines God as consisting of all positive attributes or all perfections, the greatest being is seen as consisting of all perfections. Just as the notion of infinite number entails infinitely many units, so it would seem that the notion of God entails infinitely many perfections. In this sense, these notions seem to be analogous, and it is hard to believe that this analogy escaped Leibniz's attention. Indeed, I suspect that it is precisely this analogy that gave rise to Leibniz's concerns about the traditional notion of God. And it is due to this apparent analogy that the impossibility of the greatest number poses a substantial threat to the possibility of the notion of the most perfect being.

It is worth stressing that, if it turned out that the notion of the *Ens Perfectissimum* were inconsistent, disastrous consequences would follow—not only for traditional rational theology (including Leibniz's) but also for the very foundations of Leibniz's metaphysics, and especially for his notion of possibility. According to Leibniz, the very notion of possibility presupposes the notion of a mind that conceives all relations between its simple attributes. Now God, seen as an omniscient mind, conceives of all possibilities in his understanding; and God's definition as entailing all perfections makes it possible for God to conceive all possibilities in his understanding by reflecting on his attributes.[20]

As it turns out, the texts are unambiguous in suggesting that the relations between the notions of the most perfect being and that of infinite number were indeed on

---

[19] A 6.3:518–19; DSR 77. See also A 6.3:523; DSR 83; A 6.3:512; DSR 67; for similar analogies and A 6.3:521; DSR 81.

[20] "The very possibility of things, when they do not actually exist, has a reality grounded in the divine existence: for if God should not exist, there would be no possibility, and possible things are from eternity in the ideas of the divine intellect" (GP VI:440). In 1678, Leibniz writes to Elizabeth, "[m]ais à présent, il me suffit de remarquer, que ce qui est le fondement de ma caractéristique l'est aussi de la démonstration de l'existence de Dieu" (A 2.1:437). For more on what I call the "actualist" strand in Leibniz's theory of possibility, see Nachtomy (2007a, ch. 1), and my "Modal Adventures Between Leibniz and Kant," in *The Actual and the Possible*, ed. M. Sinclair (Oxford, UK: Oxford University Press, 2017).

Leibniz's mind. In a letter to Conring (1677), Leibniz writes, "[t]hose that are more subtle opponents say that the most perfect being implies a contradiction just as the greatest number [*numerum maximum*] does" (A 2.1:325).[21]

Leibniz uses, juxtaposes, and contrasts these definitions even within single papers and notes. For example, in his Paris notes, he draws an explicit analogy between the essence of God and the essence of the greatest number. As we have seen, Leibniz's examples show that he is concerned not only with the notion of the greatest number, but also with other notions of maximal quantity (see A 6.3:520).[22]

It seems evident that Leibniz's analysis concerning the impossibility of the greatest number (as well the most rapid motion and the greatest shape) plays a role in promoting his concerns about the possibility of the most perfect being. His analysis of the former partly explains why the possibility of the latter requires a proof. Additional and decisive evidence that Leibniz is considering the notions of the most perfect being and the greatest number in this context appears in a later text:

> It is not quite assured whether an infinitely perfect being does not imply a contradiction, like the most rapid movement, the largest number and other similar notions, which are undoubtedly impossible. Mons. Des Cartes in his reply to the second objections, article two, agrees to the analogy between the most perfect Being and the greatest number, denying that this number implies a contradiction.[23] It is, however, easy to prove it. For the greatest number is the same as the number of all units. But the number of all units is the same as the number of all numbers (for any unit added to the previous ones always makes a new number). But the number of all numbers implies a contradiction, which I show thus: To any number, there is a corresponding number equal to its double. Therefore, the number of all numbers is not greater than the number of all evens, i.e., the whole is not greater than its part. (GP I:338)

---

[21] "At qui subtiliores sunt adversarii ajunt Ens perfectissumum tam *implicare* contraditionem quam numerum maximum."

[22] In the *Confessio*, Leibniz draws a similar analogy (A 6.3:139–40). He clearly has in mind the analogy between God, as consisting of all perfections, and the greatest number, as consisting of all units. As he writes, "[t]hings are not produced by the mere combination of forms in God, but along with a subject also. . . . The various results of forms, along with a subject, bring it about that particulars result. I cannot explain how things result from forms other than by analogy with the way in which numbers result from units—with this difference, that all units are homogeneous, but forms are different" (A 6.3:523; DSR 85).

[23] In his second objection to Descartes's *Meditations*, Caterus argued that humans may invent or think out the concept of the greatest being from their own resources, just as they may think the concept of the greatest number, though it is nevertheless impossible (AT 95–98; CSM 68–70).

The various points discussed here—that Leibniz considers the number of all numbers to be the same as the number of all units; that the proof that this notion involves a contradiction is based on Galileo's paradox; Leibniz's commitment to the axiom that the whole is greater than its parts; and the analogy between the most perfect being and the greatest number—all come into sharp focus in this passage. Taken together, they provide the grounds for Leibniz's problem—namely, how to account for both the possibility of the infinite being and the impossibility of infinite number.

We have seen that Leibniz investigates the notions of "infinite or maximal number" and "the most perfect Being" by comparing and contrasting them. We have also noted that their differences are no less significant than their similarity. Indeed, while Leibniz seeks to demonstrate the possibility of a most perfect being, he seeks to demonstrate the impossibility of an infinite number. Further, in Leibniz's eyes, the notion of the greatest being is a paradigmatic case of a *possible* notion while "the greatest number" provides a paradigmatic case of an *impossible* notion. And Leibniz's motivation to demonstrate the possibility of the most perfect being is part of his attempt to prove God's existence, for which he has not only theological commitments but metaphysical commitments as well.

If the notions of the greatest number and of the most perfect being were fully analogous, Leibniz would have to regard them as equally problematic (or equally unproblematic). As we have seen here, he does not. Rather, he considers the latter to be possible and the former to be impossible. Furthermore, he *believes that he has demonstrated* that the latter is possible and that the former is impossible. What then is the dissimilarity that Leibniz sees between the notion of the greatest being and that of the greatest number—a dissimilarity that convinces him to consider the one to be possible and the other to be impossible?

2.3 A "SEMANTIC" SOLUTION?

One might suggest that the answer to our question is to be found in the question itself—namely, in the difference between "the greatest *number*" and "the greatest *being*." In other words, we might attempt to address the difficulty by focusing on the difference between the sense of the concept of "number" and that of a "being." Despite their close syntactical similarity, the notions of the greatest number and of the greatest being also have an obvious (semantic) difference. Put crudely, this difference stems from the categorical gap between beings and numbers—a distinction that runs deep in Leibniz's metaphysics—and may be captured in their notions. We know that Leibniz considers numbers to be mental abstractions—*entia rationis*—rather than true beings. As he makes explicit in the Paris notes, "[n]umbers, modes, and relations are not entities" (A 6.3:463; DSR 7). Hence, "[i]t is not surprising that the number of all numbers [*numerum omnium numerorum*], all possibilities, all relations or reflections, are not distinctly understood; for they are imaginary and have nothing that corresponds to them in reality" (A 6.3:399; DSR 115).[24] Thus, in Leibniz's eyes, there is a straightforward connection between his observation that the number of all numbers is not distinctly understood and the fact that nothing corresponds to it.

Following this suggestion, the main difference Leibniz would see between these notions derives from the fact that the one is a notion of a being and the other is a notion of a nonbeing. For Leibniz, numbers are universal and divisible, and are composed by a conjunction of units (for more on this, see chapter 4). By contrast, beings are unique and indivisible agents. While beings, for Leibniz, must be active agents, numbers are understood as abstractions in the minds of agents. Note, however, that this is true of any number, and is not specific to the case of the "number of all numbers." The notion of God or of the most perfect Being, on the other hand, serves as the model of what a being is. In fact, it is considered the primary being, as well as the source of all possible beings.[25]

Unlike the notion of a number, the notion of God (and, in fact, of any true being, of which God is the most perfect) is, according to Leibniz, a notion of something that is not produced by composition of parts—that is, it is not something that is made up by composing or conjoining an infinite number of units or perfections. God (as well as any other being, for Leibniz) is not a sum of perfections. In fact, a being for Leibniz is not a composite sum at all; rather, it is an active and indivisible agent that is said to include all perfections not by being composed of them as parts. Such a unity cannot be fully defined in terms of its constituent elements and in this sense admits no parts at all.

While this adequately captures Leibniz's distinction between beings and nonbeings, it is not obvious that it also helps in resolving our question. The original problem was to compare the *concept* of the greatest being with that of the greatest number. And so, even if God himself is not a sum of perfections, it would not do to show that the *concept of God* does not involve a sum of perfections, such that it would make the concept of the most perfect being less problematic than the notion of an infinite number.

Certainly, the distinction between beings and numbers is relevant and important. Yet our question still stands: Why does the *notion* of the most perfect being, seen as consisting of infinitely many perfections, avoid the contradiction facing that of an infinite number? Let us not forget that Leibniz's strategy to prove that the greatest

---

[24] "[F]rom what has just been said it is clear enough that neither number, nor measure, nor time, inasmuch as they are only aids for the imagination, can be infinite" (A 6.3:280; cited in LLC 111). Compare this with Spinoza's letter on the infinite (12), LLC 103–16.

[25] See A 6.4:1618; LLC 307.

being is possible is to show that all simple positive perfections or attributes are compatible *inter se* and therefore may belong in one subject. Since a conjunction of all units would surely pass the test of internal compatibility, we still need to account for the fact that, according to Leibniz, the *notion* of the most perfect being is deemed possible if the *notion* of infinite number, seen as a conjunction of simple units, is deemed impossible. Since these notions seem to be very similar, Leibniz's radically different attitude toward them is puzzling. What makes him regard the notion of the infinite being as a paradigm of possibility and that of an infinite number as one of impossibility? I believe that the tensions involved in this complicated question motivate much of Leibniz's efforts in his complex approach to infinity. In particular, in the next section I suggest that struggling with this question leads him to distinguish between different senses of infinity, which are applicable to different contexts in his metaphysics.

2.4 INFINITE MAGNITUDE AND INFINITE PERFECTION

As we have seen, something is missing in the semantic solution (appealing to the difference between beings and numbers) given previously. Indeed, the definition of God as the "subject of all perfections" certainly does not exhaust Leibniz's notion of God. For Leibniz, God is also "necessarily a thinking being"[26]—an active mind whose activity involves thought and reflection.[27] Unlike numbers, ideas, and other incomplete notions, Leibnizian beings are intrinsically active: they are agents whose activity (and primitive force) cannot be fully understood in quantitative terms, as a mere sum of their attributes.

While one may rightly protest that activity and unity have nothing to do with the *concept* of God, this still suggests that, unlike the notion of number, Leibniz does not understand the notion of God (or that of the primary being) in purely quantitative terms. Rather, if the source of being is intrinsic activity and unity—which are qualitative features—then God's intrinsic activity and unity may indicate that the sense in which infinity is ascribed to his notion may not pertain to a quantitative aspect, as we have supposed thus far. In other words, this may indicate that the sense of infinity

---

[26] "God is not a Metaphysical something, imaginary, incapable of thought, will, action, as some make out, so that it would be the same as if you said that God is nature, fate, fortune, necessity, the World; but God is a Substance, Person, Mind. . . . It should be shown that God is a person or intelligent substance"; "Deus non est quiddam Metaphysicum, imaginarium, incapax cogitationis, voluntatis, actionis, qualem nonnulli faciunt, ut idem futurum sit ac si diceres Deum esse naturam, fatum, fortunam, necessitatem, Mundum, sed Deus est Substantia quaedam, Persona, Mens. . . . Ostendendum est Deum esse personam seu substantiam intelligentem" (A 6.3 :474–75; DSR 26. See also L 158; DSR 27).
[27] A 6.3:520; DSR 70.

that Leibniz ascribes to the greatest or most perfect being does not primarily pertain to magnitude or to number. According to this suggestion, there is a sense of "infinite" that has to do with being complete and perfect and that, strictly speaking, cannot be understood in quantitative terms. In other words, the notion of infinity as that which is not limited need not be fleshed out in numerical or quantitative terms. On this reading, the notion of the *Ens Perfectissimum* could be understood in this context in such a way that its perfection need not primarily be expressed as a *quantitative aspect* of infinite sum or number of perfections but, rather, as a *qualitative aspect* of completeness and perfection.

While Leibniz may not be clear about this in his Paris notes, some later texts do suggest something along this line of thought. In referring to his previous demonstrations, Leibniz articulates this point in terms of the difference between infinite magnitude and infinite perfection: "it has previously been demonstrated that the infinite in number and magnitude is neither one nor whole; but that only the infinite in perfection is one and whole" (*Deum non esse mundi animam*; A 6.4:1492). In addition, in discussing the essence of God, Leibniz notes, "[a]n infinite whole is one" (A 6.3:474; LLC 49). While Leibniz sees infinite *magnitude* as contradictory, it is clear that this is not the case for infinite *perfection*. This would suggest that he is using "infinite" in two distinct senses. He clearly thinks that the *Ens Perfectissimum* is both one and infinite. Thus it would seem that, in this case, the notion of infinity does not apply to magnitude and to number but, rather, to perfection and completeness.[28]

But before turning to the later texts, let us examine Leibniz's early rejection of Descartes's distinction between the infinite and the indefinite. This will shed further light on the early stages in which Leibniz develops his approach to infinity. As this discussion will show, different senses of infinity are already at play in Leibniz's very early thought; it is particularly instructive to note that, in spite of his disagreements with Descartes, Leibniz does agree with Descartes on the absolute and nonquantitative infinity of God.

2.5 LEIBNIZ AND DESCARTES: THE INFINITE, THE INDEFINITE, AND THE ATTITUDE TOWARD INFINITY

When Leibniz made his first steps into the scholarly world, Descartes's fame was already well established and his work was widely disseminated throughout Europe. It would have been surprising, therefore, if Leibniz, who was keen to read whatever he

---

[28] The connection between unity and infinity will be studied in some detail in chapter 4, this volume, in connection with Spinoza's view of substance as unique, indivisible, and infinite.

could, and especially of the new philosophers, had not been familiar with Descartes's work. However, as Maria Rosa Antognazza notes in her recent study of Leibniz's intellectual biography, "although [Leibniz] was obviously familiar with Descartes's philosophy, his knowledge of it up to [1675] had been basically second-hand. During the winter of 1675–76 and the spring of 1676 he plunged into a careful reading of Descartes' *Principia Philosophiae* (Amsterdam 1644), leaving after him a trail of notes."[29]

One thing that would not fail to draw Leibniz's attention was Descartes's distinction between the infinite and the indefinite. In part I, articles 26–27, of his *Principles of Philosophy*, Descartes distinguishes between the infinite and the indefinite (*infini vs indéfini*), and argues that we should not seek to comprehend the infinite, but should rather consider what we find without limits to be *indefinite* (art. 26, title).[30] Descartes further argues that, since we are finite beings, we should avoid discussing the infinite and thus avoid the paradoxes surrounding it:

> this is why we should not concern ourselves to respond to those who ask if half of an infinite line is infinite, and whether an infinite number is even or odd, and other similar things, because only those who imagine that their spirit is infinite have to examine such difficulties.[31]

Descartes further argues that we should reserve the term "infinite" for God alone, for only God's nature can be properly called infinite. All other things that we perceive to have no limits, such as the extension of the universe or the number of the stars, should be regarded as indefinitely large. Descartes further argues that their indefiniteness does not belong to their nature but, rather, stems from the fact that human understanding is limited and deficient, and therefore cannot perceive the infinite (art. 27).[32]

---

[29] Antognazza (2009, 167).
[30] For a translation of Leibniz's comments on this article, see LLC 25.
[31] "c'est pourquoi nous ne nous soucierons pas de répondre à ceux qui demandent si la moitié d'une ligne infinie est infinie, et si le nombre infini est pair ou non pair, et autres choses semblables, à cause qu'il n'y a que ceux qui s'imaginent que leur esprit est infini qui semblent devoir examiner telles difficultés" (art. 26, CSM I :201–202; AT VIIIA:14–15).
[32] "26. Qu'il ne faut point tâcher de comprendre l'infini mais seulement penser que tout ce en quoi nous ne trouvons aucunes bornes est indéfini.

Ainsi nous ne nous embarrasserons jamais dans les disputes de l'infini; d'autant qu'il serait ridicule que nous, qui sommes finis, entreprissions d'en déterminer quelque chose, et par ce moyen le supposer ni en tâchant de le comprendre; c'est pourquoi nous ne nous soucierons pas de répondre à ceux qui demandent si la moitié d'une ligne infinie est infinie, et si le nombre infini est pair ou non pair, et autres choses semblables, à cause qu'il n'y a que ceux qui

Leibniz's note on Descartes's distinction (arts. 26 and 27) reads as follows:

> Instead of "infinite," he recommends that we use the term "indefinite," i.e. that whose limits cannot be found by us, and that the term "true infinity" should be reserved to God alone. But contrary to this, in Part 2, article 36, matter is admitted to be really divided by motion into parts that are smaller than any assignable, and therefore actually infinite. (A 6.3:214; LLC 25)

It is interesting to observe that, already in his early "Theory of Abstract Motion" of 1671, Leibniz sharply objects to Descartes's distinction between the infinite and the indefinite along a similar line of reasoning—that is, by defending the actual division of the continuum. He writes,

> There are actually parts in the continuum, contrary to what the most acute Thomas White believes, and *these are actually infinite*, for Descartes's "indefinite" is not in the thing, but in the thinker. (Winter 1670–71; A 6.2:264; LLC 339)

From this note we learn that, even before he had direct access to Descartes's *Principles of Philosophy*, Leibniz criticized Descartes for grounding the distinction between the infinite and the indefinite epistemologically—that is, in human limitations for comprehending infinity. Both remarks (of 1671 and of 1675) suggest that Leibniz is not wary of endorsing the actual division of matter to infinity. Moreover, his 1675 remark suggests that he thinks that Descartes, too, endorses something

---

s'imaginent que leur esprit est infini qui semblent devoir examiner telles difficultés. Et, pour nous, en voyant des choses dans lesquelles, selon certains sens, nous ne remarquons point de limites, nous n'assurerons pas pour cela qu'elle soient infinies, mais nous les estimerons seulement indéfinies. Ainsi, parce que nous ne saurions imaginer une étendue si grande que nous ne concevions en même temps qu'il y en peut avoir une plus grande, nous dirons que l'étendue des choses possibles est indéfinie; et parce qu'on ne saurait diviser un corps en des parties si petites que chacune de ses parties ne puisse être divisée en d'autres plus petites, nous penserons que la quantité peut être divisée en des parties dont le nombre est indéfini; et parce que nous ne saurions imaginer tant d'étoiles que Dieu n'en puisse créer davantage, nous supposerons que leur nombre est indéfini, et ainsi du reste.

27. Quelle différence il y a entre indéfini et infini. Et nous appellerons ces choses indéfinies plutôt qu'infinies, afin de réserver à Dieu seul le nom d'infini; tant à cause que nous ne remarquons point de bornes en ses perfections, comme aussi à cause que nous sommes très assurés qu'il n'y en peut avoir. Pour ce qui est des autres choses, nous savons qu'elles ne sont pas ainsi absolument parfaites, parce qu'encore que nous y remarquions quelquefois des propriétés qui nous semblent n'avoir point de limites, nous ne laissons pas de connaître que cela procède du défaut de notre entendement, et non point de leur nature."

like the actual division of matter, but that he uses the evasive terminology of indefinite division so as not to acknowledge it. Note, too, that Leibniz already recasts Descartes's position in his own terms: rather than referring to the "indefinitely divisible," he uses the phrase (which I suspect he adapts from Hobbes) "smaller than any assignable"—a phrase which, for him, implies the syncategorematic sense of infinity.[33] Against Descartes's attempt to reserve the use of infinity for God alone, Leibniz would hold that there are many things other than God that can be adequately seen as (and called) infinite.

In spite of Leibniz's critique of Descartes's distinction between the infinite and the indefinite, it is important to observe that there is one issue on which they agree: for both, the infinity of God is absolute, and in the terms we are using here, for both, the infinity of God is not seen as a quantitative kind of infinity. In other words, for both Leibniz and Descartes, the infinity of God does not relate to greatness in magnitude. In a letter to Henry More, Descartes makes this point rather explicitly:

> God is the only thing I positively understand to be infinite. As to other things like the extension of the world and the number of parts into which matter is divisible, I confess I do not know whether they are absolutely infinite; I merely know that I know no end to them, and so, looking at them from my own point of view, I call them indefinite.[34]

In his second letter to More, Descartes writes:

> I say . . . that the world is indeterminate or indefinite, because I do not recognize in it any limits. But I dare not call it infinite as I perceive that God is greater than the world, not in respect to His extension, because, as I have

---

[33] The notion of "less than any given or assignable number" is articulated in Hobbes's definition of *conatus*: "conatus is motion through a space and a time less than any given, that is, less than any determined whether by exposition or assigned by number, that is, through a point" (*De corpore* 3.15.2; T. Hobbes, *The English Works of Thomas Hobbes of Malmesbury*, ed. W. Sir Molesworth, 11 vols. [London: Bohn, 1839–1845; reprinted 1966], I:177). Decisive evidence that Leibniz read *De corpore*, presented through his annotations on *De corpore*, was discovered and presented by Ursula Goldenbaum in U. Goldenbaum and D. Jesseph, eds., *Infinitesimal Differences: Controversies Between Leibniz and His Contemporaries* (Berlin and New York: Walter de Gruyter, 2008), 76–94. See also Douglas Jesseph's discussion in the same volume ("Truth in Fiction: Origins and Consequences of Leibniz's Doctrine of Infinitesimal Magnitude," in *Infinitesimal Differences* [Berlin and New York: Walter de Gruyter, 2008], 216–18); Arthur (2014, 57); LLC xxxiii; Antognazza (2009, 110); D. Garber, *Leibniz—Body, Substance, Monad* (Oxford and New York: Oxford University Press, 2009), 3, 14–16.

[34] Descartes to Henry More, February 5, 1649; CSMK 364.

already said, I do not acknowledge in God any proper [extension], but in respect to His perfection.[35]

The infinity of God, according to Descartes, relates primarily, and perhaps exclusively, to his perfection. With respect to the infinity of God, Leibniz's view is very similar to Descartes's view. God's infinity does not pertain to extension or to any magnitude or other quantitative feature; rather, God's infinity pertains exclusively to perfection. It goes without saying that, for both Descartes and Leibniz, God is defined (in accordance with the tradition) as the most perfect Being (*Ens Perfectissimum*). As Leibniz states, "The absolute is prior to the limited." "And just so the unbounded is prior to that which has a boundary [terminus], since the boundary is something added" (A 6.3:502; A 6.3:392; GP I:224). And, as Robert Adams clarifies, "Leibniz's conception of divine perfection commits him to agree with Descartes that, in its own nature, the divine infinity or perfection is primitive—that it is unanalyzable and not a negation of the finite. For him, as for Descartes, the infinite, in properties capable of infinity, is the primary case, and the finite is formed by limitation, or partial negation, of the infinite (NE 157f)" (Adams 1994, 116).

But Leibniz's agreement with Descartes ends here. For Descartes states that there is no other thing that we should qualify as infinite—that we should refrain from ascribing infinity to things which seem unbounded to us because, in the final analysis, we cannot comprehend what infinity means. Further, Descartes argues that finite beings (such as we are) should not pretend to understand, or even try to understand, something infinite. Leibniz's attitude is almost the inverse. For Leibniz, there are many things (series, worlds, individuals) that can be (and, in fact, must be) understood as infinite. However, in describing all these things as infinite, Leibniz is working with different senses of infinity. And, as noted, he agrees with Descartes that only God is infinite in the *absolute* and nonquantitative sense.

One might argue that this disagreement between Leibniz and Descartes is merely about words and that, in the end, there is no substantial difference in their positions. It is arguable that Leibniz's distinction between different senses of infinity comes down to something quite similar to Descartes's distinction between the infinite and the indefinite. Indeed, in his article, "Leibniz on the Indefinite as Infinite" (1998), Bassler argues that this is precisely the case. According to Bassler, in Leibniz's notes from 1676, one finds a distinction that is very similar to Descartes's. Bassler observes

---

[35] Descartes's second letter to Henry More, May 15, 1649, quoted by Koyré (1957, 122) (and not translated by CSMK). In a letter to Clerselier from April 23, 1649, Descartes explains, "[b]y 'infinite substance' I mean a substance which has actually infinite and immense, true and real perfections" (AT V:355; CSMK 377).

(1998, 850) that Leibniz approves of the indefinite progression of natural numbers and rejects the notion of an infinite number of (finite) numbers. This observation has some basis in the texts, but Bassler's assimilation of Leibniz's notion of (the syncategorematic) infinite with Descartes's notion of the indefinite blurs some important differences between their views. Bassler argues that, in his later work, "Leibniz takes the indefinite as infinite" (852). However, it is clear that Leibniz himself thought that his disagreement with Descartes was not merely terminological but, also, substantial.

Notice, first, that the distinction Bassler is referring to is drawn within the realm of mathematics. Here, Leibniz uses the notion of infinite number as an illustration of something impossible, for an infinite number cannot be conceived and thus has no consistent notion. Yet, Leibniz qualifies as infinite some other things, which are not impossible. An infinite series is one obvious example. Leibniz sees an infinite series as possible because he defines a series through its generation rule (or law of the series) and not as a sum of its constituents (see chapter 1).

Second, as already noted, Leibniz rejects Descartes's view that the distinction between infinite and indefinite is due to the incapability of our (finite) mind to understand the infinite. One obvious reason for this conviction is his mathematical work during his years in Paris. His work on infinite series and the calculus shows that he sees both the notion of infinitesimally small and of infinitely large as mathematically manageable and indeed very useful. But it should be noted that Leibniz's response to Descartes precedes his development of the calculus. And thus it seems clear that his early commitment to investigate the infinite does not depend on his mathematical work. In addition, Leibniz holds that, although we certainly do not fully comprehend the infinite, we can nevertheless demonstrate some things about it. This point is clearly expressed in a letter to Malebranche from 1679:

> The number of all numbers implies a contradiction, which I show thus: to any number there is a corresponding number equal to its double. Therefore the number of all numbers is not greater than the number of even numbers, i.e. the whole is not greater than its part. It is no use responding that our finite mind cannot comprehend the infinite, for we can demonstrate something about what we do not comprehend. And here we comprehend at least the impossibility, if this only means that there is a certain whole which is not greater than its part.[36]

---

[36] Leibniz to Malebranche, June 22, 1679; GP I:338, translation in G. Brown, "Leibniz's Mathematical Argument Against a Soul of the World," *British Journal for the History of Philosophy* 13, no. 3 (2005): 479. This point comes up in other passages as well: "At last a certain new and unexpected light shined from where I least expected it, namely, from mathematical

Leibniz's conclusion from this reasoning is that an infinite sum of parts, seen as a whole, is an impossible notion. But this negative result has some positive implications: it leads Leibniz, in contrast to Descartes, to make positive observations about the infinite. According to Leibniz, we can say that there are infinitely many things or parts of matter as long as we do not see them as a single whole or as a true unity. As early in 1672, Leibniz observed that "[t]here is no maximum in things, or what is the same, the infinite number of all unities is not one whole, but is comparable to nothing" (A 6.3:98; LLC 13).[37] Thus, for Leibniz, it would be misguided to reduce infinity to something that we call undefined or undetermined because we cannot comprehend it. Indeed, for Leibniz, there is no categorematic infinity of things. But the notion of infinity is extremely useful. As he notes in his piece from 1676, "Infinite Numbers":

> we conclude finally that there is no infinite multiplicity, from which it will follow that there is not an infinity of things either. Or it must be said that an infinity of things is not one whole, i.e. that there is no aggregate of them.[38]

Leibniz's conclusion here (in 1676) is that one can talk about infinitely many things as long as one does not regard these things as a totality or as making up a single whole (that would also admit of parts).[39] This is an important point that Leibniz firmly

---

considerations on the nature of infinity. For there are two labyrinths of the human mind, one concerning the composition of the continuum, and the other concerning the nature of freedom, and they arise from the same source, infinity. That same distinguished philosopher I cited a short while ago preferred to slash through both of these knots with a sword since he either could not solve the problems, or did not want to reveal his view. For in his *Principles of Philosophy* I, arts. 40–41, he says that he can easily become entangled in enormous difficulties if we try to reconcile God's preordination with freedom of the will; but, he says, we must refrain from discussing these matters, since we cannot comprehend God's nature. And also, in *Principles of Philosophy* II, art. 35, he says that we should not doubt the infinite divisibility of matter even if we cannot grasp it. But this is not satisfactory, for it is one thing for us not to comprehend something, and quite something else for us to comprehend that it is contradictory" (1689? "On Freedom"; AG 95).

See also: "having contented himself with saying that matter is actually divided into parts smaller than all those we can possibly conceive, [Descartes] warns that the things he thinks he has demonstrated ought not to be denied to exist, even if our finite mind cannot grasp how they occur. But it is one thing to explain how something occurs, and another to satisfy the objection and avoid absurdity" ("Pacidius to Philalethes," October 29–November 10, 1676; A 6.3:554; LLC 183–85).

[37] See also NE 2.17.
[38] April 10, 1676, "Infinite Numbers"; A 6.3:503; LLC 101.
[39] See also the "Conversation of Philarete and Ariste"; AG 267; GP VI:592.

holds for the rest of his career. As an example, consider this passage from Leibniz's letter to Bernoulli of 1699:

> Given infinitely many terms, it does not follow that there must be an infinitesimal term. … I concede the infinite multiplicity of terms, but this multiplicity does not constitute a number or a single whole. It signifies nothing but that there are more terms than can be designated by a number. Just so, there is a multiplicity or complex of numbers, but this multiplicity is not a number or a single whole.[40]

As we have seen earlier, Leibniz denies the possibility of infinite quantities. But this, he thinks, need not prevent us from using infinity. For to refer to infinitesimals or infinite series is not to refer to true wholes or to true entities. In other words, one need not suppose the existence of an infinitely small (or large) quantity (or entity) in order to use the infinitesimal calculus (or to apply infinity more generally).

Since Leibniz's critical response to Descartes's distinction between the infinite and the indefinite appears in early in his work (1671), it seems that Leibniz held this approach even before he started his serious work in mathematics (under the guidance of Huygens in Paris). Indeed, not only did Leibniz hold the position before developing the calculus but it would have been a partial cause for his *developing* the calculus. While such a claim seems to read much of the later development into Leibniz's early comment, Leibniz's early comment does point to some of the intuitions that might have led him to *develop* his calculus.[41]

Be this as it may, my main point here is that Leibniz's approach to investigating the infinite stands in stark contrast to Descartes's. Descartes recommends avoiding any discussion of the infinite and, especially, pretending that we can comprehend it. As Descartes writes to Mersenne,

> I have read M. Morin's book. Its main fault is that he always discusses the infinite as if he had completely mastered it and could comprehend its properties. This is an almost universal fault which I have tried carefully to avoid. (January 28, 1641, Descartes to Mersenne; AT III:293; CSMK 171–72)

---

[40] February 21, 1699, Leibniz to Bernoulli; GM III:575; L 514, with translation from Levey (1999, 139).
[41] See Arthur, in LLC xxxiii.

Descartes adds,

> I have never written about the infinite except to submit myself to it and not to determine what it is or what it is not.[42]

Leibniz's attitude toward the question of infinity could not be more different. As we shall see, unlike Descartes, Leibniz does attempt to provide a positive account of the infinite and the productive ways in which it can be used in mathematics, as well as in metaphysics.

2.6 CONCLUSION

Neither Descartes nor Leibniz is shy about defining God as having infinitely many attributes. Indeed, the same is true of Spinoza, who encounters a similar problem to Leibniz's, as we shall see in the next chapter. However, Spinoza's refusal to ascribe infinite number to God seems less problematic because, for him, questions of possibility are merely epistemic and merely betray a deficiency of knowledge. For Leibniz, however, consistent concepts indicate pure possibilities. Leibniz's possibility proof of "a most perfect being," of which he takes pride throughout the rest of his life, turns on his observation that a subject that includes all positive perfections does form a consistent concept. But it is exactly such a concept that seems to suggest an infinite number of perfections. Leibniz proves that a being whose notion consists of infinitely many attributes is consistent, but a number whose notion consists of infinitely many units is not. Just as the notion of infinite number implies infinitely many units, the notion of God seems to imply infinitely many perfections or attributes.

If this were the case, however, Leibniz would have to consider both notions to be equally problematic. Yet, he clearly does not believe this to be the case. Rather, he considers the notion of an infinite being to be possible, while he considers the notion of an infinite number to be impossible. If the notions of infinite number, most rapid motion, and the greatest shape are all contradictory, as Leibniz holds, he has to show that the notion of the infinite being is not.

Roughly stated, Leibniz's approach to the problem turns on the observation that an infinite being is infinite in a nonquantitative sense. Such a being is not a whole composed of discrete parts nor, strictly speaking, does it admit of parts at all; rather, it is said to be indivisible and immeasurable in the sense that its being is not something that can be quantified and divided.

---

[42] Descartes to Mersenne, January 28, 1641; AT III:293, translation in R. Ariew, "The Infinite in Spinoza's Philosophy," in *Spinoza: Issues and Directions*, Proceedings of the Chicago Spinoza Conference, September 1986, vol. 14, ed. E. Curley and P. F. Moreau, Studies in Intellectual History (Leiden and New York: Brill, 1990), 17.

With respect to the consistency problem, I suspect that Leibniz's solution turns on stressing that both infinite number and the notion of an infinite being are concepts. In other words, concepts are not entities; certainly, Leibniz does not regard them as entities or true beings. If his syncategorematic solution works for the notion of an infinite number, it should work for the *concept* of the most perfect being as well. While the concept should not be seen as a unity, the entity itself must be seen as a unity. The concept indicates a possibility—and thus can be regarded in a quantitative sense; the entity is seen as a real, actual being and hence has to be regarded as infinite in a qualitative sense.

That Leibniz thinks about the difference between these two notions along the lines described in this chapter is shown in his letter to Malebranche (cited earlier). He writes,

> [y]ou will tell me that there is an idea of the perfect being, since we think of this being and, therefore, it is possible. But it will be answered that by the same reason one could say that there is an idea of the greatest number and that one can think of it. We see, however, that it implies [a contradiction]. It is true that there are reasons to distinguish between these infinite impossibilities, like the [greatest] number and the [most rapid] movement and other similar things, and the supremely perfect being. But here one needs novel and rather profound reasoning to be assured of this [difference]. (GP I:339, my translation)

Leibniz, however, remains rather cryptic about this "novel and profound reasoning" about the difference between the impossibility of a greatest number, shape, and so forth and the possibility of an infinite being.

Having shown that Leibniz was preoccupied with the relations between the notion of the greatest being and that of the greatest number, I have suggested two ways in which he might have conceived of the difference between these notions—one that stresses the difference between being and number, and one that stresses the difference between two notions of infinity—namely, an infinity that pertains to quantitative features of magnitude and number, on the one hand, and an infinity that pertains to qualitative features of perfection and completeness, on the other. While these attempts may be less than satisfactory, they play an important role in Leibniz's use of infinity in his metaphysics, since they push him to distinguish between different senses of infinity. Leibniz's distinction between different senses of infinity is crucial in order to account for the difference between the notions of infinite being and infinite number. According to this suggestion, the notion of "infinite" in "infinite number" and in "infinite being" is not the same. While the former pertains to quantity, the latter pertains to being complete and perfect in a way that cannot be quantified or enumerated. In the next chapter, I will discuss Spinoza's approach to a similar problem and will present Leibniz's attempt to resolve these tensions through his response to Spinoza.[43]

---

[43] There is another way to approach this question. One could stress the dissimilarity between the units of infinite number, which are homogeneous, and God's perfections, which are not. If God's attributes are primitive to the extent that they are incommensurable, then Galileo's paradox may not arise. If God's perfections are such that they cannot be compared and related to one another as parts and whole, then it would seem that the paradox could not be generated. Indeed, in one text Leibniz says explicitly that this is the difference between units and forms. He writes: "I cannot explain how things result from forms other than by analogy with the way in which numbers result from units – with this difference, that all units are homogeneous, but forms are different" (A 6.3:523; DSR 85). While this suggestion is interesting, I cannot see why the attributes, even if entirely heterogeneous and primitive, may not be indexed and correlated with numbers.

3

LEIBNIZ READS SPINOZA
DIFFERENT SENSES AND DIFFERENT DEGREES
OF INFINITY

3.1 LEIBNIZ'S ENCOUNTER WITH SPINOZA

In 1676, Leibniz reluctantly traveled from Paris back to Hanover. He made a point to travel via The Hague in order to meet Spinoza.[1] As far as we know, the two philosophers met once. Their philosophical systems, however, have many more meeting points. In The Hague, Leibniz showed Spinoza his modified version of Anselm's (and Descartes's) proof for the existence of God. According to this version of the argument, the notion of the *Ens Perfectissimum* entails existence, since it includes all perfections, and since existence is considered a perfection. As noted in the previous chapter, Leibniz found this reasoning unsatisfactory on the following grounds: one needs to show not only that the conclusion follows from the premises but also that the definition of the *Ens Perfectissimum* is consistent—a point taken for granted by all previous upholders of the argument. In other words, Leibniz argues that, in order to prove that a most perfect being exists, one has to show first that this notion is consistent—that is, as we have seen, that it is possible (A 6.3:572; DSR 91; A 6.3:583; DSR 105–107).[2]

---

[1] Unless otherwise indicated, translations of Spinoza's works are from Curley, vol. 1. References to the letters not translated by Curley are to Gebhardt, vol. 4. In writing this chapter I have greatly benefited from Mogens Lærke, *Leibniz lecteur de Spinoza. La genèse d'une opposition complexe* (Paris: Honoré Champion, 2008).

[2] Cf. a note from 1676, in which Leibniz writes, "[i]n the chapter of St. Thomas' *Summa Contra Gentiles* which is entitled 'Whether the existence of God is known per se,' there is a reference to an elegant argument which some use to prove the existence of God. The argument is: God is that than which nothing greater can be thought. But that than which nothing greater can be thought cannot not exist. For then some other thing, which cannot not exist, would be greater than it. Therefore God cannot not exist. This argument comes to the same as one which has often been advanced by others: namely, that a most perfect being exists. St. Thomas offers a refutation of this argument, but I think that it is not to be refuted, but that it needs supplementations. For it assumes that a being which cannot not exist, and also a greatest or

Leibniz's reasons for presenting *this* particular argument to Spinoza are curious, but as we have no evidence as to Leibniz's reasons, I'd rather avoid speculating. Whatever reasons Leibniz had, this much is clear: the issue Spinoza and Leibniz discussed in their meeting in The Hague is highly indicative of some of the remarkable affinities, as well as some of the deep rifts, between their views regarding the nature of the infinite, and especially the relation between the notions of infinite being and infinite number. In fact, while Leibniz's approach implies that the existence of an infinite and most perfect being follows from its essence, Spinoza holds that, since being finite involves some negation, infinity expresses (or, one might say, is) the very absolute affirmation of existence.[3]

As I will argue in this chapter, Leibniz's understanding of numbers and his (nonquantitative) view of the infinite being are very similar to Spinoza's. At the same time, there is a subtle difference between Spinoza and Leibniz: the latter stresses that the notion of an infinite being has to be conceived as a pure concept in order to show that such a being is possible.[4] But Spinoza explicitly rejects the notion of pure logical possibility (E IP33). In other words, Leibniz demands a possibility proof, which Spinoza cannot accept precisely because the notion of pure possibility requires an abstraction from existence. This difference turns out to be crucial for the ways in which Leibniz and Spinoza conceive of the notion of an infinite substance.

In what follows, I focus on the characterization of a substance as an infinite being. We see this in Spinoza's definition of God: "a being constituted by infinitely many attributes."[5] As he states clearly in his Epistle 12, "every substance can be understood

---

most perfect being, is possible" ("On Truths, the Mind, God, and the Universe," April 15, 1676; A 6.3:510–11; DSR 63).

[3] "Since to be finite is some negation and to be infinite is an absolute affirmation of the existence of some nature, it therefore follows from proposition 7 that any substance must necessarily exist" (E IP8S). Spinoza argues further that one can adequately consider the uniqueness of being (i.e., of substance) with respect to its existence alone and not its essence. As Spinoza writes to Jarig Jelles, "in an *Appendix to the Principles of Descartes, Geometrically Demonstrated* I established that God can be called one [*unum*] or unique [*uniqum*] only in a very inappropriate sense, I respond that a thing cannot be called one and unique with respect to essence but only with respect to existence. We conceive of things as existing in a certain number of exemplars only if they are brought under a common genus" (Ep. 50 to Jarig Jelles; Gebhardt IV:239). I take this to imply that, according to Spinoza, one cannot *conceive* of the unique and infinite being in abstraction from its existence, as a pure essence. In addition, one may talk about the unique existing thing in a numerical sense only in an inappropriate sense. For the category of number can only apply to things that can be "brought under a common genus," which obviously does not hold of a unique being.

[4] "God is a being from whose possibility (or, from whose essence) his existence follows. If a God defined in this way is possible, it follows that he exists" (A 6.3:582; DSR 105).

[5] Spinoza's definition of God reads as follows: "Per Deum intelligo ens absolute infinitum, hoc est, substantiam constantem infinitis attributis, quorum unumquodque infinitam et aeternam essentiam exprimit" (E IP6D).

only as infinite" (A 6.3:277; LLC 105; Curley 202), and "if we attend to it [substance] as it is in the intellect, and perceive the thing as it is in itself . . . then we find it to be infinite, indivisible, and unique" (A 6.3:277; LLC 105; Curley 203). Leibniz, too, holds that infinity, indivisibility, and uniqueness are essential characterizations of substance. As noted, however, according to Leibniz, the concept of such a being must be conceived as a pure logical possibility, as a precondition for an argument for its existence.

A striking difference in their metaphysical systems is that, for Spinoza, there is only a single substance, whereas Leibniz speaks of infinitely many beings. As it is usually put, Spinoza is a substance monist and Leibniz is a substance pluralist. Indeed, this distinction captures a major difference between Spinoza's and Leibniz's metaphysical systems. Yet a close analysis reveals some fundamental agreements regarding the claim that any substance is, by definition, both infinite and unique. In light of this similarity, I will suggest that, when Leibniz and Spinoza say that the divine substance is infinite, it is to be understood primarily in a nonquantitative sense.[6]

In the previous chapter I argued that Leibniz was preoccupied with the difference between the notion of infinite number, which he regards as impossible, and that of the infinite being, which he regards as possible. In the following section I examine Spinoza's solution to a similar problem, which he expounds mainly in his letter on the infinite (Ep. 12). Leibniz read and annotated this letter in April 1676.[7] The gist of what I call Spinoza's solution is to distinguish between various kinds of infinity and, in particular, between one that applies to substance and one that applies to numbers, seen as auxiliaries of the imagination. Leibniz, I argue, accepts this kind of approach and adapts it to his own use. But, as typical of him, Leibniz does not simply borrow or accept Spinoza's solution. Rather, he modifies and tailors it to his own terminology and his own ends. Leibniz recasts Spinoza's distinctions between different *types* of infinity (A 6.3:282; LLC 114–15) in terms of *degrees* of infinity. These degrees are (1) *omnia* (absolute infinity), which applies to God alone; (2) *omnia sui generis*, or maximum in its own kind; and (3) *infinitum tantum*, or mere infinity, which applies to numbers and other *entia rationis* (in a syncategorematic sense).

---

[6] The thesis of the nonquantitative sense of infinity, as presented here, also provides a partial explanation for why Leibniz was attracted to Spinoza's philosophy during his years in Paris (especially in 1675–76) and, at the same time, why he ultimately moved away from it while reading and commenting on Spinoza's *Ethics* in 1678. The chapter thus tells a dual story, one conceptual and one historical. For details of the complex way in which Leibniz read Spinoza, see Lærke (2008).

[7] See Lærke (2008, 423–24). This letter is not the only source Leibniz obtains regarding Spinoza's views at the time. He receives quite accurate information on the *Ethics* from Tschirnhaus, with whom he discusses Spinoza's metaphysics, as well as questions of mathematics (see, for example, Leibniz's letter of May 1678; GM IV:451–63; L 294–99).

This threefold distinction, I suggest, is consistent with that between a nonquantitative concept of infinity (ascribed to the divine substance) and a quantitative concept of infinity (ascribed to numbers and magnitudes). On the basis of this threefold distinction, I will suggest in the following section that Spinoza and Leibniz hold a similar, nonquantitative conception of the infinity of substance. I observe that this conception of infinity surfaces in Leibniz's reading notes for Spinoza's *Ethics*, in 1678.

3.2 SPINOZA'S SOLUTION: LETTER 12 AND LEIBNIZ'S ANNOTATIONS

Like Leibniz, Spinoza too has to account for the difference between the infinity of magnitudes and the infinity of God. Spinoza explicitly defines God as "a being absolutely infinite, that is, a substance consisting of an infinity of attributes" (E ID6).[8] In Epistle 12, Spinoza takes the following approach to this problem: he distinguishes between different kinds of infinity and, in particular, between a kind of infinity that applies to (indivisible) substance and a kind of infinity that applies to divisible quantities. This approach emerges explicitly when Spinoza discusses the nature of the infinite and how to dissolve the traditional paradoxes surrounding it.

Spinoza's argumentation in this letter is of considerable complexity. However, one point is rather clear: Spinoza holds that the notion of infinity that may apply to the substance is nonquantitative. Since Spinoza identifies God with the one substance, God's infinity is not comparable to that of numbers. The reason is that any reference to numbers presupposes a limitation and hence would imply that it is finite.[9] According to Spinoza, the tendency to describe a substance with numerical infinity is entirely misguided and generates contradictions. The way out of the contradictions affecting the infinite is to avoid the common confusion between the *quantitative sense* of infinity that can adequately be ascribed to number and divisible quantities, and the sense of infinity that can adequately be ascribed to a unique and indivisible substance.

---

[8] The translation is significant here. It can also be translated "consisting of infinite attributes." This is important in identifying the kind of infinity that is at work here. In addition, this plays into the debate regarding the number of attributes that God is said to have. Those who translate it as "an infinity" tend to hold that there is a numeric infinity of attributes, while "infinite attributes" is related to the infinite nature of the attributes. I prefer the former, but have opted to using Curley's translation throughout. I thank Noa Shein for a discussion of this note.

[9] "From what has just been said it is clear enough that neither number, nor measure, nor time, inasmuch as they are aids for the imagination, can be infinite. For otherwise number would not be number, nor measure, nor time" (LLC 111).

It is worth noting that Leibniz begins his annotations on Spinoza's letter by stating that Spinoza "demonstrates that every substance is infinite, indivisible, and unique" (A 6.3:275; LLC 101). Then, Leibniz copies (almost to the letter, although with some significant modifications) Spinoza's definitions of substance (E ID3) and of God (E ID6). This certainly tells us something about the interest Leibniz takes in reading this letter. Particularly indicative here is Leibniz's addition to Spinoza's definition of God.

> *He defines God as follows:* that which is an *absolutely* infinite being, i.e. a substance consisting of infinite attributes, each of which expresses an infinite and eternal essence and is thus immense [*immensum*]. (LLC 103)[10]

The clause "adeoque immensum est" is nowhere to be found in Spinoza's definition; it is, rather, added by Leibniz. This is telling. First, *immensum* is not a term used by Spinoza. More important, in his annotations to this letter (A 6.3:282; L 24; LLC 115), Leibniz states the following: "I have always distinguished the *Immensum* from the *Interminato*, i.e., that which has no bound [*seu terminum non habente*]." In his notes from this period, Leibniz is using *Immensum* as a noun—the *Immensum*— designating God as something beyond measure. He also uses *Immensum* as "that which persists during continuous change in space . . . and is one and indivisible" (A 6.3:519; see LLC 450). Evidently, Leibniz is using the notion of *Immensum* in more than one sense. Likewise, divine immensity is taken as the "basis of space" (A 6.3:519; LLC 450). However, it seems clear that, unlike the current English connotations of the word "immense," Leibniz does not use *immensum* here to indicate immense *magnitude*; rather, he uses it in a way much closer to its literal meaning in Latin—that is, to indicate something *beyond any measure* or, more precisely, something that has no measure (and is therefore impossible to measure)—something that cannot be measured because it does not belong to the category of quantity.

Both recent English translators, Parkinson and Arthur, have emphasized this point (see DSR 122n92 and LLC 450). To avoid the current English connotations of "immense," Parkinson renders *immensum* as "immeasurable," so that the Latin negation of measure (*mensura*) would remain conspicuous in the translation. As Arthur notes in the glossary for his edition, "*Immensum* can be synonymous with 'infinite' or 'beyond measure,' as Leibniz employs it in Aiii4: 95; and at Aiii60: 475, where Leibniz distinguishes this species of the infinite from the unbounded" (LLC 450, Latin–English glossary).

---

[10] "*Deum sic definit.* Quod sit Ens *absolute* infinitum, hoc est substantia constans infinitis attributis, quorum unumquodque infinitam et aeternam essentiam exprimit adeoque immensum est" (A 6.3:276).

Thus, when Leibniz adds his gloss to Spinoza's definition of God—that is, that the absolutely infinite being is also *immensum*—he refers to one of his own notions of infinity, viz., that which is beyond measure. In this sense, *immensum* is distinguished from the unbounded. The unbounded infinite designates a *measurable* quantity, whereas *immensum* designates something that *cannot be measured*. Thus, Leibniz intends to emphasize that God is something beyond any measure—something that cannot be described in quantitative or measurable terms and is inadequately described in quantitative terms.[11] To recap, the main point here is that Leibniz's addition indicates that, in his eyes, the infinity of the divine substance cannot be quantified or measured but, rather, belongs to an altogether different category.[12]

Leibniz then adds a very interesting note on a being conceived through itself (*per se concipi*):

> something is *understood through itself* only if we conceive all its requisites without having conceived another thing, i.e., only if it is the reason for its own existence. For we commonly say that we *understand* things when we can *conceive* their generation, i.e., the way in which they are produced. Hence we understand through itself only that which is its own cause, i.e., that which is necessary, i.e., is a being in itself. And so it can be concluded from this that if we understood a necessary being, we would understand it through itself. But it can be doubted whether we do understand a necessary being, or, indeed, whether it could be understood [*intelligatur*] even if it were known or recognized [*cognosci*]. (A 6.3:275, LLC 101)

In reading Spinoza's letter, Leibniz seems to recall the difficulty of showing that the notion of a necessary being can be understood or, in other words, that it is intelligible. According to Leibniz, in order to show that something is intelligible, one has to show how it is produced by giving a real definition in the sense we explored in the previous chapter—that is, by showing that its concept is consistent. Thus, it seems that, in reading Spinoza's letter, Leibniz is still occupied with his own problem as well.

Given Leibniz's preoccupation with the tension between the possibility of an infinite being and the impossibility of an infinite number, it is not surprising that he is interested in the way that Spinoza connects the definitions of substance,

---

[11] Note that this sense of *immensum* as being beyond any measure applies to Leibniz's usage of *immensum* as "the basis of space" as well. While Leibniz is not using Spinoza's terminology, his gloss of Spinoza's notion of God seems to capture Spinoza's notion of God as infinitely extended quite well.

[12] For a slightly different emphasis on Leibniz's addition of the word *immensum*, see Lærke (2008, 469–77 and 424–25).

God, and infinity. He agrees with Spinoza that any substance "is infinite, indivisible, and unique." Yet, according to him, the possibility of such a being needs to be demonstrated. While he maintains his demand to prove the *possibility* of the most perfect being, Leibniz does not restate (but only mentions) his proof after 1676.[13] As we saw in the previous chapter, Leibniz demonstrates that an infinite collection of discrete units is impossible and cannot be regarded as a whole. At the same time, he clearly regards God as an infinite unity. In fact, he calls God the one-all (*unus omnia*; A 6.3:385), and maintains that such a being is possible.

In light of this, one can reasonably suppose that Leibniz would seek support for his line of reasoning regarding the possibility of an infinite being and the impossibility of an infinite number. Such support may indeed be found in Spinoza's letter. Toward the beginning, Spinoza notes that,

> everyone has always found the problem of the Infinite very difficult. Indeed insoluble.[14] This is because they have not distinguished between what is infinite as a consequence of its own nature, or by the force of its definition, and what has no bounds, not indeed by the force of its essence, but by the force of its cause. And also because they have not distinguished between what is called infinite because it has no limits and that whose parts we cannot explain or equate with any number, though we know its maximum and minimum. Finally, they have not distinguished between what we can only understand, but not imagine, and what we can also imagine. (LLC 103; Curley 201)

He adds:

> If they have attended to these distinctions, I maintain that they would never have been overwhelmed by such a great crowd of difficulties. For then they

---

[13] Adams (1994) has argued convincingly that the a priori proof for the possibility of the notion of the *Ens Perfectissimum* gives way to a presumption of its possibility. See chapter 8, this volume, and A 2.1:436 for an explicit text endorsing the presumption of possibility. Lærke (2008) notes that, after 1677, Leibniz only mentions his a priori proof but it never appears in his later writings. I am not sure what conclusion should be drawn from this. It is obvious that Leibniz maintains that the *Ens Perfectissimum* is possible. However, it is not obvious on what grounds. From the fact that he does not repeat the argument, it cannot be concluded that he abandoned it, as Lærke seems to hold. The presumption of possibility might be an addition to, rather than a replacement for, the a priori argument. As far as I can see, we simply cannot tell.

[14] In his exposition of Descartes's *Principles of Philosophy*, Spinoza mentions some of the traditional difficulties associated with the infinity: "if an infinite is not greater than another, quantity A will be equal to its double, which is absurd"; "whether half an infinite number is also infinite, whether it is even or odd, and the like" (Gebhardt I:190; also see 192–96); Also see Ariew (1990).

would have understood clearly what kind of Infinite cannot be divided into any parts, or cannot have any parts, and what kind of Infinite can, on the other hand, be divided into parts without contradiction. They would also have understood what kind of Infinite can be conceived to be greater than another Infinite, without any contradiction, and what kind cannot be so conceived. (LLC 103–105; Curley 202)

According to Spinoza, there are different kinds of infinity that correspond to different kinds of things (viz., substance, attributes, and modes).[15] By means of these distinctions, Spinoza qualifies and restricts the way in which infinity can be ascribed to substance. Spinoza's distinction suggests an attractive approach to Leibniz's problem. According to Spinoza, one kind of infinity (the one pertaining to infinite being) "cannot be divided into any parts, or cannot have any parts," and the other kind of infinity (the one pertaining to modes) "can . . . be so divided into parts without contradiction." This is of course related to Spinoza's view that, strictly speaking, a substance is infinite and indivisible (E IP15S). For this reason, a substance is not divided into parts but, rather, its attributes have various modes.

According to Spinoza, the kind of infinity that we can ascribe to substance is such that "we cannot explain or equate with any number." An infinite substance, on this view, is nondivisible and cannot be understood in numerical terms. For this reason, the use of this kind of infinity would not involve the contradictions that affect things whose enumeration requires comparison and abstraction by the imagination. In fact, Spinoza maintains that enumeration involves abstraction and comparison of things under a common genus by means of the imagination.[16] However, the kind of infinity that pertains to a substance cannot even be adequately conceived by the imagination but only by the intellect. This can clearly be seen in the following passage from Epistle 12 (which summarizes E IP15S):

we conceive quantity in two ways: either abstractly, or superficially, as we have it in the imagination with the aid of the senses; or as substance, which is done by the intellect alone. So if we attend to quantity as it is in the imagination, which is what we do most often and most easily, we find it to be divisible, finite, composed of parts, and one of many. But if we attend to it as it is in the intellect, and perceive the thing as it is in itself, which is very difficult, then we

---

[15] Duration, number, and motion are seen as mere auxiliaries of the imagination, which serve as measures of divisible magnitudes. Cf. M. Gueroult, *Spinoza I: Dieu* (Paris: Aubier-Montaigne, 1968; Ariew (1990); and Lærke (2008).

[16] See E IP15, Ep. 34, Ep. 50, and the next section for more details.

find it to be infinite, indivisible and unique, as I have already demonstrated sufficiently to you before now. (A 6.3:278; LLC 107; Curley 202–203)[17]

This last point—which Leibniz mentions in the first line of his annotations to Spinoza's letter—suggests a way out of the inconsistency he identifies in infinite quantity. In line with Spinoza's reasoning, Leibniz can distinguish between "beings" and "nonbeings" by observing that each requires a different kind of infinity. And this would account for his regarding an infinite being as possible and an infinite number as impossible. At the same time, the concept of an "infinite being" is to be taken in a syncategorematic sense (or more precisely, the term "infinite" in the concept of an infinite being).

Thus, what is most pertinent for Leibniz's purposes in Spinoza's letter can be paraphrased as follows: any number is by definition limited. For this reason, it is also measurable. By contrast, God's infinity cannot be quantified, measured, or numbered, precisely because this would imply limiting it (or seeing it as limited), as well as viewing God as a divisible and discrete entity, which Spinoza clearly regards as absurd. This suggests that, for Spinoza, "infinity" is used differently when ascribed to numbers (or more generally to divisible and discrete quantities, or to a feature of modes and abstractions) and when it applies to the all-inclusive substance or God. A substance is said to be infinite on account of its completeness and absolute perfection. Therefore, for Spinoza, it must be *in*divisible and admit of no parts. In this sense, a substance is said to be infinite in a nonquantitative sense.

Given the context presented in the first section, it should now become clear why Leibniz would be receptive to such a view. Indeed, he seems to agree with Spinoza's analysis. Yet, as is typical of him, Leibniz recasts Spinoza's distinction in his own terms and appropriates it for his own purposes.[18] In his annotations, he writes,

> I set in order of degree: *Omnia; Maximum; Infinitum*. Whatever contains *everything* is maximum in entity; just as a space unbounded in every direction is maximum in extension. Likewise, that which contains everything is most infinite [*infinitissimum*], as I am accustomed to call it, or the absolutely infinite.

---

[17] Cf. "If therefore we consider quantity as it is in the imagination, that which is the most ordinary, we find that it is finite, divisible and composed of parts; if, on the contrary, we consider it as it is in the understanding and we conceive it insofar as it is substance, then, as we sufficiently demonstrated, we will find it to be infinite, unique and indivisible" (E IP15S).

[18] Obviously, I do not argue here for a direct influence in the sense that Spinoza's letter is the exclusive or even the main source for Leibniz's views on infinity. Rather, I claim that Leibniz's attraction to Spinoza's view is evident in his annotations and that his response to Spinoza's views is revealing of and serves him to articulate his own views.

The *Maximum* is everything of its kind, i.e., that to which nothing can be added, for instance, a line unbounded on both sides, which is obviously also infinite; for it contains every length. Finally those things are *infinite in the lowest degree* whose magnitude is greater than we can expound by an assignable ratio to sensible things, even though there exists something greater than these things. . . . For a maximum does not apply in the case of numbers. (A 6.3:282; LLC 114–15)[19]

As Lærke notes,

If one compares this classification with the one proposed in Spinoza's letter 12, one is struck by their similarity. First, the distinction between *maximum* and *omnia* evokes the distinction between the attributes, which are infinite "in their kind" in EIdef.4 and the "absolutely infinite" substance in EIdef.6—which is exactly the definition reproduced at the beginning of the *Communicata ex litters domini Schulleri*. [Likewise] there is a strong resemblance between that which Leibniz calls "*immensum*" and that which Spinoza calls "infinite by nature." (Lærke 2008, 433, my translation)

The similarity between *immensum* and "infinite by nature" is particularly remarkable. As Lærke also notes, "that which is infinite by nature or by virtue of its definition, is the substance" (2008, 430). We have already seen that Leibniz amends Spinoza's definition of God with the clause "that which is *immensum*."

There are, however, some significant differences between Leibniz and Spinoza here. The most conspicuous difference is that Leibniz reformulates Spinoza's distinction in terms of degrees. *Omnia*, he says, "is the highest degree, [it] is everything, and this kind of infinite is God, since he is all one; for in him are contained the requisites for existing of all the others" (A 6.3:385; LLC, 43). Elsewhere, and later in his career, Leibniz is also very clear that the highest degree, the "absolutely infinite," applies to God alone. For example, in a letter to Des Bosses, from March 11, 1706, he notes that "only indivisible and absolute infinite has true unity: it is God" (GP II:305). In the *New Essays* (NE 2.17.1), he writes that "rigorously speaking, the true infinite is only in the absolute, which is anterior to any composition and is not formed by the addition of parts" (GP V:144, my translation).[20] This notion of absolute infinity is

---

[19] Compare with A 6.3:385; LLC 43, where Leibniz articulates the same threefold distinction in slightly different words.
[20] "Le vrai infini à la rigueur n'est que dans l'absolu, qui est antérieur à toute composition, et n'est point formé par l'addition des parties" (GP V:144; NE 2.17.1).

nonquantitative in the sense that God or the most perfect Being has a nondivisible unity, which admits of no parts; also, it cannot be compared to or measured by any quantity. In this sense, absolute infinity indicated allness and perfection, which cannot be measured. So this notion of infinity is aptly called the Immeasurable or *Immensum*. It involves absolute perfection, completeness and, most important to my concerns here, inherent unity and indivisibility.

As we have already noted, this conception of the infinite could help Leibniz in avoiding the difficulty facing the notions of infinite number, line, speed, shape, or any other magnitude. Simply stated, on such a conception of infinity, quantitative categories are inapplicable to true beings.[21] And likewise, maximal quantities cannot be regarded as perfections ("Discourse on Metaphysics," paragraph 1). Therefore, if infinity is ascribed to a substance not in a quantitative sense but only in the sense of absolute infinity, the notion of infinite substance or being, qualified in this way, would avoid the contradiction of infinite number and other infinite magnitudes.

Leibniz thus reserves the notion of the absolutely infinite for God, or the most perfect Being. While the quantitative infinite does not pertain to a complete being (which is also a unity), infinity in the absolute sense of *Omnia*, does. Even though their views about the divine substance are quite different, the connection between infinity and unity is crucial for both Spinoza and Leibniz. In this regard, Leibniz and Spinoza share the following view: substance is the only thing of which one can say that it is an infinite and unique being. As we have just seen, however, this conception involves a nonquantitative understanding of the infinity (and, possibly, of the uniqueness) of substance. In the next section, I examine the extent to which Leibniz and Spinoza share this conception.

3.3 LEIBNIZ'S COMMENTS ON SPINOZA'S ETHICS: ON NUMBERS AND THE INFINITY OF BEING

In 1678, once the *Opera posthuma* was published, Leibniz could finally read Spinoza's *Ethics*. This reading results in some interesting comments. Of particular interest for our purposes is his comment on proposition 8 of part I. In the second scholium to this proposition, Spinoza writes,

> I. . . . the true definition of each thing involves nothing and expresses nothing but the nature of the thing defined.

---

[21] This is at least true in the case of God. The case of created substances is much more delicate.

From which it follows,

II. that no definition involves or expresses any certain number of individuals, since it expresses nothing else than the nature of the thing defined. For example, the definition of a triangle expresses nothing else than the simple nature of a triangle, but not any certain number of triangles.

It is to be noted,

III. that there must be, for each existing thing, a certain cause on account of which it exists.

Finally, it is to be noted,

IV. that this cause, on account of which a thing exists, either must be contained in the very nature and definition of the existing thing (*viz., that it pertains to its nature to exist*), or must be outside it.

From these propositions it follows that if, in Nature, a certain number of individuals exists, there must be a cause why those individuals, and why neither more nor fewer, exist.

For example, if twenty men exist in Nature... it will not be enough... to show the cause of human nature in general; but it will be necessary in addition to show why not more and not fewer than twenty exist. For (by III) there must necessarily be a cause why each [NS: particular man] exists. But this cause (by II and III) cannot be contained in human nature itself, since the true definition of man does not involve the number 20. So (by IV) the cause why these twenty men exist, and consequently, why each of them exists, must necessarily be outside each of them.

For that reason it is to be inferred absolutely that whatever is of such a nature that there can be many individuals [of that nature] must, to exist, have an external cause to exist. (Curley 414–15)

Spinoza argues here that a substance cannot have its reason or cause from any external thing but must constitute its own reason or cause, for "it pertains to its nature to exist." If all numerical determinations must have an external cause, any numerical ascription could not apply to the substance. Given this, a substance cannot be enumerated or understood in numerical terms; for any ascription of number will have nothing to do with its essence. According to Spinoza, number is a purely extrinsic denominator that cannot apply to substance as part of its definition or essence (E IP15S; Ep. 34).

While Leibniz's attitude in his comments on Spinoza *Ethics* in 1678 is highly critical, he calls this argument "elegant" (A 6.4:1770). As Lærke notes, when Leibniz says of an argument that it is elegant, it is probably because it resembles his own (Lærke 2008, 673). Indeed, Leibniz's view regarding the status of numbers as abstractions from a plurality of different individual things closely resembles Spinoza's view as expressed in the argument just quoted . For example, Leibniz writes,

> When I perceive a horse and an ox, I note that the ox is not *the same*, but *different*. But since they combine in something there will be *many* things, to wit, animals or beings. But that which can be substituted for another without altering the truth is *the same*. (1683; A 6.4:561; LLC 267)

According to Leibniz here, the judgment that there are *many* things stems from observing differences and similarities: the horse and the ox are the same insofar as we regard them under a common genus, that is, as animals. However, they differ in the kind of animal they are. Leibniz goes on to exemplify his point in a more general way as follows:

> But if A is D, and B is D, and C is D, and A, B, and C are the same, D will be *one* thing. If, on the other hand, A, B, and C are each different from the other, they will be *many*, whence numbers. (A 6.4:561; LLC 267)[22]

The point I would like to emphasize here is that numerical ascriptions derive from observing a multiplicity of different things. According to Leibniz, numbers, like relations, are not entities in their own right (A 6.3:463; DSR 7).[23] Rather, numbers depend on mental abstractions: they are products of comparisons between particular things. When Leibniz says in the prior passage "whence numbers," he means that numbers arise by observing such similarities and differences among particular things. Compare this view with Spinoza's point in his Epistle 50 to Jarig Jelles:

> We do not conceive things as existing in a certain number, unless we have reduced them to a common genus. For example, one who holds in his hand a penny and a crown will not think of the number two, unless he can call both the penny and the crown piece by one and the same name, to wit, coins or pieces of money. In the latter case he can say that he holds two coins or pieces of money, inasmuch as he calls the crown as well as the penny, a coin, or a piece of money. (Gebhardt IV:239–40)

---

[22] For additional references to Leibniz's later view on the ideal nature of number, on a par with relations and possibilities, see GP II:268–69, 276–79, 282; GP IV:568.

[23] This is consistent, of course, with Leibniz's well-known quasi-nominalist approach to abstract concepts, *possibilia* and relations. See M. Mugnai, "Leibniz's Nominalism and the Reality of Ideas in the Mind of God," in *Mathesis rationis*, ed. Albert Heinekamp, Wolfgang Lenzen, and Martin Schneider (Munich: Nodus, 1990), 153–67.

The similarity between Leibniz and Spinoza's approach to the nature of numbers is evident here. To make this point even more clear, one can observe that, in a letter to Sophie from 1700, Leibniz restates this point by quoting the Duke of Bourgogne in a passage that echoes Spinoza's example from the scholium to *Ethics* I, proposition 15:

> when one attentively considers the existence of beings... one understands very clearly that existence belongs to unities, and not to numbers (or multitudes). Twenty men only exist because each man exists. Number is nothing but the repetition of the unities to which existence only belongs. There would never be any number, if there weren't some Unites. (GP VII:560)

Leibniz comments on this: "I have read all this with admiration, and I find that my ideas concerning unities are wonderfully well expressed [*merveilleusmement bien exprimée*]" (GP VII:560). What Leibniz finds here so wonderfully well expressed seems to be that unities exist, whereas numbers are "nothing but the repetition of the unities." These, in turn, arise from comparing and considering particular things under a common genus (cf. A 6.3:399; DSR 115).

According to this conception of numbers, the status of (the number) 1 is particularly interesting. Cleary, the number 1 cannot be formed by grouping individual entities or by comparing such entities to any other particular thing. For such operations presuppose a multiplicity of entities. It seems, therefore, that one should be regarded as a limit case or as a unit that constitutes numbers (the so-called element of number) rather than a number itself. This latter treatment of the number 1 is well grounded in a tradition stretching back to Aristotle, and Leibniz's understanding of numbers does presuppose basic units or individual things. For this reason, I suspect that when Leibniz, following Spinoza, says of a substance that it is one, he is not necessarily making a numerical claim; it may well be that he is, rather, pointing to the basic unity and uniqueness of the substance. This view and some of its implications in Leibniz's metaphysics is further explored in the next chapter.[24]

---

[24] When it is said that "X is one," at least three distinct things might be meant: X is a unit; X is unique; X is one in number, comparable to 2, 3, 4, etc. As I expand in the next chapter, I think that a nonnumerical definition of substance applies in the two interesting (limit) cases of "one" and "infinity," which for both Leibniz and Spinoza are important characteristics of a substance. Both uniqueness and infinity, I suggest, are often used nonnumerically in qualifying a substance; in many contexts, what Leibniz intends to stress is that a substance is a unit and that it is unique. This does not rule out, of course, the possibility of saying that there are so many substances or even counting them in certain contexts, as Leibniz's comment on Spinoza's E IP8 indicates. The question of unity and uniqueness is examined in detail in the next chapter.

With this view of numerical ascription in mind, let us now return to Leibniz's comment on Spinoza's *Ethics,* part I, proposition 8. Leibniz reformulates Spinoza's argument as follows:

> Given several individuals, there has to be a reason in nature for this number [of individuals] rather than another. The same reason which explains why there are such a number of them must also explain why this or that individual exists. But this reason is no more in the one [individual] than in the other. Hence, it is external to all of them. (A 6.4:1770, my translation)[25]

As far as I can see, this is a very close paraphrase of Spinoza's point.[26] After reformulating Spinoza's point, however, Leibniz turns to raise an objection: "One could object by saying that their [the individuals'] number is unbounded (*interminatum*), or null (*nullum*), or exceeds any number" (A 6.4:1770).[27] Leibniz's objection indicates that he considers various options regarding the numerical status of individuals here. He notes three alternatives: (1) The number of individuals might be unbounded (*interminatum*); (2) it might be null—that is, it might admit of no numerical value at all;[28] and (3) it might exceed any number, which suggests that it is to be read syncategorematically—that is, as a variable that can be taken as small or large as desired (see section 1.3). This gives us a sense of the range of options Leibniz considers with respect to the application of numbers to individuals. His objection is very instructive, not only because it shows the way in which Leibniz thinks about this question but also because it shows that he does not intend to refute Spinoza's argument (which he in fact endorses). As Leibniz observes, "one can take several among them [the individuals] and consider why *they* exist; or consider several [of them] which have something in common such as to exist in a certain time or place" (A 6.4:1770). Thus, once again, Leibniz endorses here the idea that a numerical

---

[25] "quia ponantur esse plura individua, ideo debet esse ratio in natura cur sint tot non plura. Eadem cum faciat cur sint tot, faciat cur sit hoc et hoc. Ergo et cur sit hoc. Ea ratio autem non est in uno horum potius quam in altero. Ergo extra omnia" (A 6.4:1770).

[26] However, the application of the principle of sufficient reason is quite different in Leibniz and Spinoza. For Leibniz, there is a thinking agent, namely God, who actually executes these considerations with a view for choosing the best of all possibilities, whereas Spinoza rules out the notion of nonexisting possibilities (E IP29 and 33).

[27] "Una objectio fieri posset, si dicatur numerum [eorum] esse interminatum, sive nullum, sive omnem numerum excedere" (A 6.4:1770).

[28] It seems to me that part of Leibniz's point in calling "infinite magnitudes" nothing (*null* or *nihil*) is to stress that they are not to be seen as very small or very large but as admitting of no measure at all—only finite measures would make sense. See Leibniz's notes on Galileo's *Two New Sciences* in Fall 1672, where Leibniz compares infinite number to nothing (*nihil*) (A 6.3:168; LLC 9).

ascription is based on a common denominator of any (definite or indefinite) multiplicity of things and that a numerical ascription, therefore, is not part of the essence of these individual things.

If a numerical ascription is not part of the essence of things but derives from observing some of their common properties, and if infinity and uniqueness are essential features of substance, then it would follow that the characterization of substance as infinite should not be seen as a *numerical* qualification. This suggests that the infinity ascribed to God in saying that he is infinite in perfection, power, and knowledge is to be seen as qualitative rather than quantitative. Likewise, the notion of perfection as constituting that of the most perfect being or of an infinite being in Spinoza is not quantitative, either.[29]

As Leibniz clarifies in the first paragraph of the "Discourse on Metaphysics," the notion of perfection is one that does not admit of a maximum. For if it did, it would be contradictory, just as the notion of an infinite number is contradictory. We may conclude that, for Leibniz, both the notion of perfections, seen as attributes of the most perfect being, and the most perfect being itself are seen nonquantitively. This would indicate, as Leibniz says in his correspondence with Eckhard (A 2.1:543; GP I:266; L 177), that perfection is the highest degree of reality or being.[30] But the highest degree of perfection and infinity is not something that can be measured or understood in terms of quantity; rather, it is, as in Leibniz's addition to Spinoza definition of God, *immensum*—that is, something beyond measure in that ascribing measure to it would involve the categorical mistake of trying to quantify something that is only qualitative.

3.4 CONCLUSION

A close study of Leibniz's texts before and after his encounter with Spinoza between 1675 and 1679—first through the mediation of Tchirnhaus, then face to face in The Hague, and then in Leibniz's reading of the *Ethics* in 1678—reveals some points of close similarity in their treatment of the infinity of substance. As we have seen, in the texts from 1675–76, Leibniz's view concerning the infinity of substance is very close to Spinoza's. Moreover, Spinoza's view on infinity offers an attractive way to approach Leibniz's problem and clarify his approach regarding the way in which a substance

---

[29] Gueroult has made this point with respect to Spinoza: "the assertion that each attribute expresses an infinite and eternal essence simply signifies that each attribute expresses a thing which is infinite by its nature (or essence), that is, poses its existence absolutely (E IP8S) and is eternal, i.e., exists necessarily by itself (definition 8)?" (Gueroult 1968, 68).
[30] This point is developed in section 8.1, this volume.

may be said to be infinite. This approach turns on the distinction between two senses of infinity: one that can be quantified and one that cannot.

But, in fact, Leibniz's reading of Spinoza reveals more than this. He recasts Spinoza's distinction between kinds of infinity, each with a different domain of application, in terms of degrees ("I set in order of degree: *Omnia; Maximum; Infinitum*"). Roughly speaking, between the highest degree of infinity, which Leibniz clearly ascribes to the absolute and necessary being, and the lowest degree of infinity, which he ascribes to *entia rationis* such as numbers and relations, Leibniz invokes a third, intermediate degree of infinity: a maximum in its kind.

In chapter 8, I will suggest that Leibniz's intermediate degree of infinity is useful in order to characterize the nature of *living* beings. Unlike the tradition, and in distinction to all major thinkers at the time, who accepted a sharp dichotomy between an infinite creator and finite creatures, Leibniz takes created beings to be infinite in important respects (while still being finite and limited in many other respects). Unlike God, created beings are not absolutely infinite; but they are infinite in the sense that each created substance may be seen as a complete entity in its own kind. As we know well, created beings, for Leibniz, are individuals. A created being may be seen as infinite in the sense of a maximum in its own kind (*maximum in suo scilicet genere*). I now turn to examine more closely the connections between infinity and unity.

4

INFINITY AND UNITY
MATHEMATICS AND METAPHYSICS

And it would be the subject of a fine inquiry to discover whether, out of the entire totality of finite numbers, one number can be defined which is the most beautiful of all – unless perhaps this is the number one, which represents all power at the same time.
—G. W. Leibniz (A 6.3:477; DSR 31)[1]

There are no substances where there is no substance, just as there are no numbers where there are no unities; [b]ut just as all numbers are derived from one plus one, so must all multiplicity be derived from unity.
—G. W. Leibniz, Fardella memo, 1672[2]

4.1 INFINITY AND ONENESS, OR ONE IS SAID IN MANY WAYS

According to both Spinoza and Leibniz, a substance—and God as such—is *one*. For both thinkers, as for most thinkers at the early modern period, God is considered to be both one and *infinite*. In the *Ethics*, part I, proposition 14, corollary C, Spinoza writes, "from this it follows most clearly, first, that God is unique, i.e., (by D6), that in Nature there is only one substance, and that it is absolutely infinite (as we indicated in P10S)" (Curley 420). In Epistle 12, Spinoza states, "every substance can be understood only as infinite" (A 6.3:277; LLC 105; Curley 202), and "if we attend to it [substance] as it is in the intellect, and perceive the thing as it is in itself . . . then we find it to be infinite, indivisible, and unique" (A 6.3:277; LLC 105; Curley 203). In the same letter, Spinoza writes,

---

[1] "Et disquisitionis foret egregiae, an ex tota numerorum finitorum universitate aliquis definiri posit, omnium pulcherrimus, nisi is forte unitas, quae simul omnes potenti refert" (*De arcanis sublimium vel de summa rerum*, February 11, 1676; A 6.3:477; DSR 31).
[2] A 6:4; LBr 389; translation in A. P. Coudert, *Leibniz and the Kabbalah* (Dordrecht: Kluwer Academic, 1995), 83. See also Letter to Sophie, GP VII:557.

What I should like you to consider about substance are: (1) that existence pertains to its essence, i.e. that from its essence and definition alone, it follows that it exists (if I am not mistaken, I have demonstrated this to you before in conversation, without the aid of other propositions); and (2) (which follows from (1)) that substance is not manifold; rather, there exists only one unique substance of the same nature; and Finally, (3) that every substance can be understood only as infinite. (Ep. 12; LLC 105)

As we have seen in the previous chapter, Leibniz agrees with Spinoza that infinity, indivisibility, and uniqueness are defining features of every substance. When Leibniz reads Spinoza's letter in 1676, the first thing he notes is that "[Spinoza] demonstrates that every substance is infinite, indivisible, and unique" (A 6.3:275; LLC 101).

Indeed, traditional theology typically treats God as a unique and infinite substance. This view of God holds true for the nontraditional metaphysics of both Spinoza and Leibniz, as well. But the conjunction of oneness and infinity in a single substance requires some clarification. If both these features are understood in a quantitative sense, they are incompatible.[3] In other words, if the notion of God is considered as both one and infinite, then either oneness or infinity (or both) must be held in a nonnumerical sense. In the previous chapter I have discussed the distinction between quantitative and non-quantitative senses of the notion of infinity in Spinoza and Leibniz. In this chapter, I shall focus on a similar distinction between different senses of oneness. I attempt to show that Leibniz's qualification of a substance as "one being" is intended to emphasize the essential unity and indivisibility rather than a numerical feature of substance.[4]

Let us first distinguish among three different senses of oneness that may be said of a substance:

(i) A substance may be one in number. In this sense, we may say of a substance that it is one thing as opposed to two, three, or four such things;

---

[3] From a traditional point of view, there is nothing remarkable in saying that God is both one and infinite. Rather, these are two of the attributes ascribed to God at least since Augustine. It goes without saying that Spinoza does not consider unity and infinity to be attributes in the strict sense of the term. Compare also Galileo's point about the relation between infinity and unity: "Infinity and indivisibility are in their very nature incomprehensible to us; imagine then what they are when combined. Yet if we wish to build up a line out of indivisible points, we must take an infinite number of them, and are, therefore, bound to understand both the infinite and the indivisible at the same time" (G. Galileo, *Dialogues Concerning Two New Sciences*, trans. H. Crew and A. de Salvio [New York: Dover, 1914], 30; EN 77–78).

[4] In this connection, see H. Ishiguro, "Unity Without Simplicity," *The Monist* 81 (1998): 534–52.

(ii) A substance may be one in the sense of uniqueness, so that there is nothing exactly like it;

(iii) A substance may be one in the sense of a unity. Here it is useful to distinguish a weak sense of unity—that is, that of parts held together or united into one whole, and a strong sense of unity, which amounts to indivisibility or even a stronger sense, the lacking of parts. Leibniz famously distinguishes between *unum per se* and *unum per accidens*,[5] or substantial unity vs. unity by aggregation. Whereas unity by aggregation implies divisibility, substantial unity implies indivisibility.[6] In his later writings, Leibniz stresses the strongest sense of unity when substances are characterized as lacking parts. Daniel Garber has emphasized that "substance as such must lack parts; it must be a genuine unity and not just an aggregate that is united."[7]

These three senses of oneness—oneness in number, uniqueness, and unity—are closely and interestingly related; but they are certainly not identical.[8] As we have just seen in the case of unity, each of these senses might admit of further distinctions. For my purposes here, the rough differences just mentioned should suffice. For what I seek to show here is that Leibniz's qualification of a substance as "one being" is primarily intended to emphasize the essential unity and indivisibility of a substance. My claim can also be expressed by noting that unity per se (or an indivisible unity) implies numerical oneness but not vice versa.

In the following section (4.2), I discuss Spinoza's and Leibniz's views of the relation between numbers and substances and show that the term "one" figures not as a number but, rather, as a fundamental unity presupposed in two contexts—that is, as a unit presupposed by any plurality and also as a unit presupposed by any of its fractions. With this conception of one as a basic unit in mind, I turn (in sections 4.3 and 4.4) to examine the metaphysical context, which, as I shall argue against a recent

---

[5] Letter to Arnauld from April 30, 1687.

[6] These distinctions are also noted by Lærke. He writes, "I will focus on the three terms which, in Spinoza's texts, are immediately presented as having a bearing on whether the cardinal number 'one' can be assigned to God in any proper manner. These terms are *unity, uniqueness,* and *oneness*" (M. Lærke, "Spinoza's Monism? What Monism?," in *Spinoza on Monism*, ed. P. Goff [Hampshire: Palgrave Macmillan, 2012], 246).

[7] Garber (2009, 330). But Garber also locates the moment in which the "lack of parts" becomes an essential characteristic of substance (rather than the indivisibility in the correspondence with Arnauld) in a draft leading to the "New System" (Garber 2009, 325).

[8] See Aquinas's distinction between one as the negation of division viz., as implying metaphysical unity and one as the principle of number (belonging to the genus of quantity). See *Summa Theologiae* I.11.a.1 (Aquinas, *Summa Theologiae, Questions on God*, ed. Brian Davies and Brian Leftow [Cambridge: Cambridge University Press, 2006], 105–106).

reading of Levey, is an essential part of Leibniz's qualification of a substance as one being. In section 4.3, I address Leibniz's famous statement in the letter to Arnauld: "that which is not one being is not a being." I argue that Leibniz wishes to emphasize the sense of unity, so that unity, in the sense of indivisibility, becomes essential for articulating the nature of substance. This reading is further confirmed by the reciprocity of being and unity that Leibniz articulates in the same letter to Arnauld, as well as in his early letters to Des Bosses by using the scholastic formula *ens et unum convertuntur*. In section 4.4, I discuss various analogies between mathematical and metaphysical unity, and consider in particular the law of the series as a Leibnizian strategy to reconcile infinity and unity in his view of substance.

4.2 SPINOZA AND LEIBNIZ ON NUMBERS AND SUBSTANCE(S)

As we have seen in the previous chapter, Spinoza's and Leibniz's views concerning the nature of numbers are similar. The judgment that there are *many* things is based on regarding *different* particular things under the same aspect.[9] Moreover, Spinoza maintains (Ep. 12) that numbers are auxiliaries of the imagination: they arise when certain exemplars are seen under a common genus. In this picture, numerical claims do not pertain to the essence of the things being numbered. In other words, numerical claims add nothing to the nature of particular things, since they are mere abstractions that are based on properties they share. As we have also seen in the previous chapter, Leibniz's view regarding the status of numbers is practically the same.[10]

According to Leibniz, the judgment that there are many things stems from observing differences and similarities: the horse and the ox are the same insofar as they are regarded as animals, but they differ in the kind of animal they are. Numerical ascriptions derive from observing a multiplicity of different things under a common aspect. On this view, numbers, like modes and relations, are not entities in their own right (A 6.3:463; DSR 7).[11] Rather, numbers depend on, and result from,

---

[9] See Ep. 50 to Jarig Jelles (Gebhardt IV:239–40). In a letter to Sophie from 1700, Leibniz restates this point (GP VII:560). See section 3.3, this volume, for the full citations.

[10] 1683; A 6.4:561; LLC 267, already cited in the previous chapter. For additional references to Leibniz's later view on the ideal nature of number, on a par with relations and possibilities, see GP II:268–69, 276–79, 282; GP IV:568.

[11] As McRae puts it, numbers are extrinsic denominations, "indifferent to the things that can be enumerated, and as such are beings of the imagination" (R. McRae, "The Theory of Knowledge," in *The Cambridge Companion to Leibniz*, ed. N. Jolley [Cambridge: Cambridge University Press, 1995], 184). This is in line with Leibniz's well-known quasi-nominalist approach to abstract concepts, possibilia, and relations.

mental abstractions: they are products of the comparison between particular things. The similarity between Spinoza's and Leibniz's views on numbers is evident in these passages.

Notice that the status of the term "one" in this view of numbers, which applies to both Leibniz and Spinoza, is peculiar. Clearly, it cannot be formed by grouping individual entities or by comparing such entities to any other particular thing. Any grouping operation presupposes a multiplicity of things to be grouped. In this context, it is easier to grasp the reason why one should not be regarded as a number. Rather than a number, one is seen as the element or principle of numbers, or as the unit presupposed by any number. In a letter to Louis Bourguet from 1715, Leibniz writes: "it is true that the concept of numbers is finally resolvable into the concept of unity, which is not further analyzable and can be considered the primitive number" (L 664).

The idea that one is the foundation of all numbers is obviously in the background of Leibniz's early definition of numbers: "Number is defined as one, and one, and one, etc., or as unities" (Leibniz to Thomasius, 1669; GP I:24).[12] Leibniz would later realize that this definition applies to whole numbers alone (NE 2.16.4).[13] But the most important point for our purposes here is that Leibniz *uses* the notion of one, or *unum*, in his definition of number. This clearly shows that *unum*, according to him, is not regarded in this context as a number (for otherwise the definition would be circular). Rather, he takes the notion of *unum* to indicate any unit whatsoever, whose combination and

---

[12] "Numerum definio unum, et unum, et unum, etc., seu unitates." This idea has a long history, which goes back to Aristotle and later involves influential thinkers such as Maimonides and Aquinas. See Aristotle 1052b23–24, where unity is regarded as the principle of number. Aquinas writes, "[w]e do not predicate of God the unity with which number begins. We predicate this only of material things. For the unity with which number begins belongs to the genus of mathematical entities, which exist in matter but are defined without reference to matter" (*Summa Theologiae* Iq11a 3, in Aquinas [2006, 110]). In this connection, see also the following passage from the Jewish writer A. Harera: "As one is prior to all numbers and is pure and simple, unique in itself and contains with no other but itself all numbers, in such a way that none of them would exist and could exist without it, while it is and exists without any of them and causes them all to be by giving them their reality, their existence, and the perfection they are capable of having, and it is also found in all of them—not only in the whole but also in each of their parts—just so, and even more so, infinity (EINSOF), the first and uncaused cause, is before all things and prior to them not only in duration, which cannot be bounded but also in degree and beyond comparison" (Rabbai A. C. Harera, *Beit Elohim Vesharey Shamaim* [*The House of God and the Gate to Heaven*], ed. and trans. Yosha Nissim [Jerusalem: Yad Ben Ztvi, 2002], I:279–80; III, 3:339–38, and 4:342–41).

[13] Compare with Hobbes's definition of number: "NUMBER is one and one or one and one, and so forwards: namely, one and one make the number two, and one and one the number three; so are all other numbers made; which is all one as if one would say, number is unities" ("On Body," part II, ch. 7, §7, in Hobbes 1966, I:96).

reiteration would yield a number—that is, a plurality of units. Leibniz's approach is evident in many texts where the number 2 is defined as 1 + 1, 3 is defined as 2 + 1, 4 as 3 + 1, and so on. Thus, fundamental unities are presupposed by the notion of number, which is nothing but *a number of* any such units.

As Lærke observes, this traditional analysis of the foundational role of oneness is exactly what Spinoza denies.[14] If there were a foundational number in Spinoza, it would be 2 rather than 1, since 1 is conceivable only on the condition of the conceivability of another 1. Spinoza for his part rejects that multiplicity has a basis in any "primitive number" or "unity." Rather, for Spinoza, unity is conditioned on (the conceivability) of multiplicity—which is the exact reverse of what Leibniz holds (endorsing as he does a traditional position on the origin of numbers).

If Leibniz's view of one as the foundation of any number is indeed traditional, let us observe that this is just one aspect of a more radical thesis. For he does not only regard 1 as the element composing numbers, and thus as the unit presupposed by any plurality; he also regards 1 as a mathematical unit that is logically *prior* to any of its fractions (which may be thought of as its "internal" constituents, e.g., as 1 could be thought to consist of ½ + ½). According to Leibniz, however, fractions are not the constitutive parts of a unity; rather, they are an artificial product of its potential division. Thus, a mathematical unit remains for Leibniz most fundamental even if it can be (mentally and potentially) divided—that is, even if it is divisible.[15] For example, in a letter to Louis Bourguet from August 5, 1714, Leibniz writes,

> [w]hen I say that unity is not further analyzable, I mean that it cannot have parts whose concept is simpler than it. Unity is divisible but not resolvable, for fractions, which are parts of unity, have less simple concepts than whole numbers, which are less simple than unity, since whole number always enters into the concepts of fractions. (L 664–65)

For Leibniz, one is a basic unit that could be well divided into fractions, but could not be broken down into parts. A division of 1 into two (½ + ½) is a *mathematical* operation that concerns ideal entities. Thus we can see that Leibniz considers 1 as a fundamental unity required for both producing arithmetical multiplicity (by the addition of 1s) and for producing an internal complexity and multiplicity of fractions (by a division of 1 or by dividing 1 by other numbers).

---

[14] Lærke (2012).
[15] See Russell's discussion on this point in B. Russell, *A Critical Exposition of the Philosophy of Leibniz* (London: Routledge, 1992), 111.

This conception of 1 as the fundamental unity presupposed by any numerical plurality suggests an interesting analogy between mathematics and metaphysics. In particular, I would like to highlight here the analogy between the role oneness plays in mathematics (as a unit constituting the foundation not only of whole numbers but also of rational numbers) and the role that basic unities play in Leibniz's metaphysics.[16] Of course, Leibniz is using the comparison to stress the *difference* between indivisibility in real things and divisibility in ideal things. But I would also suggest that the status of the basic units of reality, presupposed by any plurality, is similar to that of the basic unit in the context of numbers. As mentioned earlier, for Leibniz, 1 is considered as the foundation of numbers rather than itself a number. By analogy, the essential feature of Leibniz's true units in his metaphysics, I suggest, is not oneness in number but in *unity*, which is seen as a *constitutive element of reality*. As Leibniz later put it, "actual things are composed as a number is composed from unities" (letter to De Volder, January 19, 1706; LDV 333).

4.3 *ENS ET UNUM CONVERTUNTUR*

As Fichant notes, in the correspondence with Arnauld, the substantial form functions as the operator of the substantiality of a body; and this is due to its indivisibility (Fichant 2004, 88). It is this feature—indivisibility—that makes the claim "[*v*]*oilà réellement un être*" (GP II:77; LR 146) true. As Leibniz writes, "the substance of bodies, if they are to have one, must be indivisible" (GP II:73; LR 142). Fichant argues that, in the correspondence with Arnauld, Leibniz's notion of "complete being" from article VII of the "Discourse on Metaphysics" is reformulated in terms of a complete indivisible being (*être accompli indivisible*), which is required by substantial unity. In this period, Fichant further argues, the indivisibility of a substantial form becomes the vector of *substantial unity*, which until then Leibniz had expressed by the logical unity of the (complete) notion in the "Discourse on Metaphysics" (Fichant 2004, 87). As Leibniz writes, "[a] substance requires true unity ... and I cannot conceive of any reality without a true unity."[17]

Given this central role that unity plays in Leibniz's metaphysics in this period, let us turn to examining what Leibniz "holds as an axiom" in his letter to Arnauld from April 30, 1687—namely, "ce qui n'est pas véritablement un être, n'est pas non plus

---

[16] Scholars have indeed widely discussed the ideal nature of numbers and the distinction Leibniz sketches between numbers and real things, as well as between the potential divisibility of ideal things and the actual division of real ones. But the full range of this analogy of the role of 1 has not received due attention.

[17] "Un substance demande une véritable unité" (GP II:96; LR 165); "je ne conçois nulle réalité sans une véritable unité" (GP:97; LR 165).

véritablement un être" (LR 165; AG 86; GP II:97). Notice first that the second occurrence of "un" in this phrase serves as an indefinite article rather than as an indication of the number 1. In order to emphasize this point, one might translate the second "un être" as "a being" rather than "one being"; thus, "that which is not truly *one* being is also not truly *a* being." This translation leaves out the charm of the French in repeating the same words and changing the meaning with the emphasis on *un* in the first and on *être* in the second. But it makes the point.[18] Leibniz's usage of the first *un*, I suggest, is meant to stress the essential unity of substance rather than its oneness in number.

This reading of oneness as implying the essential unity of substance relates of course to Leibniz's distinction between *the unity per se*, which he ascribes to a true substance, and *a unity by aggregation*, which plays a major role in the second part of his correspondence with Arnauld. This distinction employs the third sense of oneness (noted earlier), as an indivisible unit. As Leibniz writes, substantial unity "requires a thoroughly indivisible and naturally indestructible being" (Leibniz to Arnauld, November 28, 1686; AG 79; GP II:76).[19] Clearly, it is indivisibility that marks here a substance in distinction from an aggregate. In his letter from April 3, Leibniz argues that beings by aggregation are distinct from substances (AG 86), in that aggregates are lacking true and persisting unity. He makes the same claim a few lines earlier in the letter: "*I believe that where there are only beings by aggregation, there aren't any real beings.* For every being by aggregation presupposes beings endowed with real unity, because every being derives its reality only from the reality

---

[18] Ariew's and Garber's translation is: "*what is not truly* one *being is not truly one* being *either*" (AG 86; GP II:97).

[19] The emphasis on unity is common. Garber (2009, 325) and R. C. Sleigh (*Leibniz & Arnauld: A Commentary on Their Correspondence* [New Haven, CT, and London: Yale University Press, 1990], 104) emphasize that in his correspondence with Arnauld, Leibniz understands indivisibility as the most significant feature of substantial unity. See also P. Lodge, "Leibniz's Close Encounter with Cartesianism in the Correspondence with De Volder, in *Leibniz and His Correspondents*, ed. P. Lodge (New York: Cambridge University Press, 2004), 179; Look and Rutherford's introduction to LDB (xl); and J. K. McDonough, "Leibniz's Conciliatory Account of Substance," *Philosophers' Imprint* 3, no. 6 (2013): 9; among others. Elsewhere, Look notes: "if anything deserves to be called the fundamental tenet of Leibniz's metaphysics, it is certainly the principle of the reciprocity of unity and being. As Leibniz puts it in a letter to Arnauld in 1687, 'I hold this identical proposition, differentiated only by the emphasis to be an axiom, *that what is not truly* one *being is not truly one* being *either.*' ... That is, for something to be a real being, it must possess genuine unity. This axiom or tenet leads to the strong distinction within Leibniz's system between, on the one hand, true unities, or unities per se, and, on the other hand, aggregates, or unities by aggregation, which can have only 'phenomenal unity'" (Look 2004, 239).

of those beings of which it is composed" (AG 85; GP II:96).²⁰ But note that this point applies to aggregates.

An aggregate may be regarded as a unity only insofar as a mind observes the relations between its constituents. The unity of an aggregate is merely "a phenomenal unity or a unity of thought, which is not enough to constitute what is real in phenomena" (letter to Arnauld, October 9, 1687; WFP 132). While both a substance and an aggregate can be said to be one in number, only a substance is an *unum per se*—an indivisible and indestructible being. As in the case of one (understood as the principle of any number), it is the indivisibility of substance that is presupposed by the existence of aggregates. As we noted, Leibniz regards 1 as the foundation of any number. Analogously, he regards the basic unities of substance as the foundation of any plurality. As he notes in the Fardella memo: "there are no substances where there is no substance, no more than there would be numbers if there were no unities" (A 6.4:1672). And, "just as all numbers are derived from one plus one, so must all multiplicity be derived from unity" (LBr 389; Coudert 1995, 83).²¹

Given this view of substance as an indivisible unit, we can better understand the subsequent assertions that Leibniz makes in the letter concerning the reciprocity of one (*unum*) and being, and his dictum that "the plural presupposes the singular." Leibniz writes,

> It has always been thought that one and being are reciprocal things [*l'un et l'être sont des chose réciproque*].²² Being is one thing and beings are another; but the plural presupposes the singular, and where there is no being still less will there be several beings. What could be clearer? (AG 86; GP II:97)

However clear it seems to Leibniz, his claim can be read in more than one way. In his "On Unity, Borrowed Reality and Multitude in Leibniz,"²³ Samuel Levey challenges a reading of this passage that emphasizes the substance's substantial unity. According to Levey, Leibniz employs three arguments, and each conveys a distinct conception

---

²⁰ Cf. Paul Lodge's introduction to the Leibniz–De Volder correspondence: "Ultimately the reality of any given aggregate must be derived from some aggregate with *aggregata* that are unities, that is, things whose unity is intrinsic, which are not themselves the product of aggregation and, hence, as he puts it in Letter 55, things that 'cannot be divided into parts'" (LDV lxxxi).
²¹ See also letter to Sophie (GP VII:557).
²² Ariew and Garber translate this as "mutually supporting" (AG 86).
²³ S. Levey, "On Unity, Borrowed Reality and Multitude in Leibniz," *Leibniz Review* 22 (2012): 97–134.

of unity (2012, 97).[24] The "multitude argument," Levey holds, states that "if there are many things (what Leibniz sometimes calls 'multitude') then there must also be beings which are single things, or, to introduce yet another use of the term, unities" (2012, 113). On Levey's reading, the multitude argument rests on two premises: (a) the reciprocity of being and unity; and (b) the assertion that the plural presupposes the singular. Levey reformulates (a) as follows: "what is truly one being is truly one being, *neither more nor less*" (2003, 263; my emphasis). Levey's analysis thus endorses the numerical sense of "one" with the consequent claim that even the reciprocity of being and one "makes no special demand on the nature of being in question—unity is not understood to imply simplicity or indivisibility, for example" (2003, 262). Thus, for Levey, the question of "what it takes for something to be a single thing—some one thing" remains "a question for further metaphysical debate" (2012, 115). He writes that, "the notion of unity here just is that of a single thing, as opposed to many things, something that can be a value of a singular variable of quantification, so to speak" (2012, 114).

It seems to me, however, that Leibniz's distinction between substances and aggregates presupposes the indivisibility of substances (AG 86). At the very least, even if (b) may involve numerical implications, it still *requires* unity in the sense of indivisibility to do the metaphysical work Leibniz assigns to it—that is, to ground reality and distinguish between true beings and beings by aggregation. The metaphysical roles Leibniz assigns to unity come up in passages such as this:

> since every multitude presupposes *true unities*, it is obvious that these unities cannot be material, otherwise they would, again, be multitudes, and not true and pure unities, as are needed to make up a multitude. And thus the unities are substances apart, which are not divisible, nor, as a consequence, perishable, since everything which is divisible has parts that one can distinguish there before separating them. (Letter to Sophie, June 12, 1700; A 1.18:113–14; translation in Garber 2009, 342)

As we can see in this passage, true unities, for Leibniz, are nonmaterial, indivisible, and nonperishable—all of which are required to ground the reality of substance. Thus, I am not convinced that Levey's reconstruction of the argument provides a compelling reason for stripping "the plural presupposes the singular" from its metaphysical implications. If this were to be the case, the singular would involve no

---

[24] In his article "On Unity: Leibniz-Arnauld Revisited" (*Philosophical Topics* 3 [2003]: 262), Levey refers to this argument under the heading "the unity-and-plurality argument."

essential unity in the sense of indivisibility, persistence, and immateriality—rather heavy metaphysical work to rest on the shoulders of these indivisible units.[25]

This, I believe, is the role of the reciprocity of unity and being. As Leibniz also argues in a letter to Arnauld dated October 9, 1687: "I argue that there cannot be a plurality of beings where there is not one being, and that all multiplicity presupposes a unity" (WFP 131).

> A fraction of an animal, or a half-animal, therefore, is not one being per se, since this can be understood only of the body of the animal, which is not one being per se but an aggregate, and has an arithmetical unity and not a metaphysical unity. (LDB 31)

The reciprocity of being and unity also come to the fore in Leibniz's correspondence with Des Bosses. In the first letters written early in 1706, Leibniz refers to the scholastic formula: *"Ens et unum convertuntur."* The latter formulation suggests that *unum* indicates unity in this context. Leibniz stresses that these terms may be used interchangeably, so that *unum* could be replaced with *Ens* and vice versa. Given Leibniz's principle that two terms may be substituted only if the truth of the statement is preserved, he surely does not mean that "being" and oneness in number can be substituted and used interchangeably. In this respect, Leibniz holds that "being" and "number" are not only distinct but also *opposed*. As we have seen here, Leibniz maintains that a number is a not a true being, but only a being of reason. Surely *unum* in its numerical sense could not be substituted with "being" in statements such as "a being truly exists" unless *unum* is used to stress an indivisible unity. While any plurality presupposes single units, the units themselves need not be regarded as one in number. Instead, they should be regarded in the foundational sense, as essential units presupposed by any plurality. In this context, too, I suggest, Leibniz uses *unum* mainly to stress the essential unity and the indivisibility of substance.

In his letter from March 2, 1706, Des Bosses complains that if being and "one" are convertible, "then nothing will exist in reality simply and actually except what is actually and simply one . . . fractions of unity and of any simple thing will be only mathematical beings that result from a mental abstraction" (LDB 25–27). In a letter dated March 11, 1706, Leibniz responds that, in saying that "being and one are convertible," he was thinking of substances that have a *metaphysical kind of unity*. The

---

[25] In the context of discussing "the indivisible, i.e., perfect, monad" endowed with primitive active force, Leibniz reaffirms that "if there is nothing that is truly one, then every true thing will be eliminated" (LDV 263).

metaphysical kind of unity is distinguished from arithmetical unity, which applies to aggregates (LDB 31). In the case of "one by aggregation," Leibniz writes, the unity in question is "semi-mental" (LDB 35).

If substances are seen as indivisible unities, there is no difficulty in counting a finite subset of them. This, however, does not make Leibniz's characterization of substance as *unum* numerical. The countability of substances only implies that the definition of substance is *consistent* with the existence of a plurality of substances and that their countability presupposes some more fundamental unity. This refers us back to the initial distinction between different senses of oneness, asking whether it could serve to mark the difference between Leibniz's and Spinoza's metaphysical systems. I argued that, while Spinoza's hesitation to refer to a substance as one relates to the relativity of numerical oneness, Leibniz's approval of oneness should not be understood on similar grounds. Instead of relating to his metaphysical pluralism, oneness relates to Leibniz's distinction between a substance and an aggregate. In addition, this understanding of oneness in terms of unity is important for Leibniz's reconciliation of the tension between oneness and infinity, a tension that arises at different levels of his metaphysics. Oneness relates not only to the infinite God but also to Leibniz's idea of one substance as involving infinitely many other substances, as well as to the infinite series of states that any simple substance involves.

4.4 ARITHMETICAL UNITY, METAPHYSICAL UNITY, AND THE LAW OF THE SERIES

The previous section discussed the notion of unity presupposed by external multiplicity and the associated distinction between an aggregate and a substance. In this section, I focus on the idea of unity presupposed by the internal complexity of Leibnizian substances. If Leibniz considers 1 as the foundation of all numbers rather than as a number, perhaps something similar would hold regarding the basic units presupposed by any multiplicity in the metaphysical context (at least in his middle and later period). Indeed, let us consider such an analogy between the mathematical and metaphysical contexts. Since a unit in mathematics may be both combined (to produce numbers by addition or reiteration) and divided (to produce fractions by division), this might have an interesting analogy in Leibniz's metaphysical context: a basic unit can be seen as the foundation of higher complex structures and as itself admitting of a complex structure. This suggests that the internal complexity of substances presupposes unity as well—not only at the bottom, as it were, but also at the top. This kind of unity, however, is not that of a plurality of units as the basis of composition (as one is the foundation of any number) but as a unity of a whole—a whole that has rich internal states and perceptions, but whose unity nonetheless may be considered in the strong sense of lacking parts.[26]

Let us begin by observing how Leibniz conceives of the internal complexity of a mathematical unit. Although a mathematical unit can be divided into fractions, it is not, according to Leibniz, reducible to (or made up of) them. An example of this reasoning comes up in a letter to De Volder from June 30, 1704:

> From the fact that mathematical body cannot be resolved into primary constituents, it may be inferred that it is certainly not real, but something mental, designating nothing other than the possibility of parts, not something actual. Indeed, a mathematical line is like an arithmetical unity, and in both cases the parts are only possible and absolutely indefinite. And a line is no more an aggregate of the lines into which it can be cut up, than a unity is an aggregate of the fractions into which it can be broken up. (LDV 303)

This idea is expressed in Leibniz's letter to Louis Bourguet from August 5, 1714 (part of which is cited earlier). In this passage he writes,

> [w]hen I say that unity is not further analyzable I mean that it cannot have parts whose concept is simpler than it. Unity is divisible but not resolvable, for fractions, which are parts of unity, have less simple concepts than whole numbers, which are less simple than unity, since whole numbers always enter into the concept of fractions. Many who have philosophized about the point and about unity in mathematics have become confused by failing to distinguish between analysis into concepts and division into parts. Parts are not always simpler than wholes, though they are always less than the whole. (L 664–65)

These passages imply that fractions presuppose unity. Whole units are simpler in the sense that fractions are produced by an operation (division) on such units (e.g., one into two halves, three thirds, etc.). However, while fractions are the products of dividing unities, the unit itself is not a product of their combination. A unit may be subject to analysis, but not subject to real disintegration into parts. A unit can be divided into parts, such as two halves, but Leibniz is clear that it is not made up of

---

[26] In the *Monadologie*, Leibniz is clear that the simplicity of the monads is not in conflict with their internal multiplicity. See, for instance, his mention of "the infinite multitude of simple substances" (*Monadologie* 57; AG 220). I discuss this feature some more in section 8.2, this volume.

them. Rather, the fraction ½ *presupposes* a unity (which is divided into two). This is the case because the notion of half already presupposes that of a unity but not vice versa. Consequently, fractions are not the constitutive parts of a unit.[27]

Note also that, in this passage, Leibniz does not focus on the ideal nature of arithmetical unities or of mathematical magnitudes in general. Rather, the main point is that mathematical unity is not resolvable into its constitutive elements because those elements are *conceptually dependent upon unity* (and not because they are indeterminate). The important point for our purposes here is that *a unity is conceptually prior to the infinity of elements that it could contain*; but does not result from combining them. This is a conceptual priority of unity over the infinite multitude and complexity that it might involve.

This analysis goes back to Leibniz's early idea that the very composition of concepts (which belong to the essences of all things) is governed by natural order—that is, an order that proceeds from the simple to the complex. Leibniz presupposes a well-defined notion of order, starting from the simplest forms and proceeding to more and more complex structures of them. This is what he calls natural order, or natural priority. According to Leibniz, natural priority derives from simplicity.[28] On my reading, the natural priority that Leibniz presupposes is due to his supposition of the priority of simples in the formation of concepts in God's mind.

Leibniz takes the priority of the simple over the complex to be evident in the context of composition. That Leibniz interprets the notion of simplicity in the context of composition is clear in the following passage: "Prior by nature is a term consisting of terms less derived. A term less derived is equivalent to one [which includes] a smallest number of primitive simple terms" (A 6.4:286–87).[29] As Rauzy remarks, natural order constitutes a general matrix to which one can refer in considering the order of things rather than the order of human discoveries.[30] The various senses of

---

[27] This last point has been discussed by commentators (such as Arthur 2001; D. A. Anapolitanos (*Leibniz: Representation, Continuity and the Spatiotemporal* [Dordrecht: Klower Academic, 1999]); T. Crockett ("Continuity in Leibniz's Mature Metaphysics," *Philosophical Studies* 94 [1999]); Levey (1999); and R. T. W. Arthur ("Leibniz on Continuity," in *PSA: Proceedings of the Biennial Meeting of the Philosophy of Science Association* (Pittsburgh: Philosophy of Science Association, 1986) in the context of the ideal nature of mathematical magnitudes and in opposition to the determinate division of other phenomena (in which unities are prior to multitudes).

[28] As he writes, "natura prius est involutum simplicius" (A 6.4:998).

[29] "Sed si sic definias: Natura prior est Terminus qui constat ex terminis minus derivatis. Terminus autem minus derivatus est, qui paucioribus simplicibus primitivis aequivalet" (A 6.4 :286–87).

[30] J. B. Rauzy, "Quid sit Natura Prius? La conception leibnizienne de l'ordre," *Revue de Métaphysique et de Morale* 1 (1995): 40.

order and various modes of priority and posteriority can be explicated by the notion of natural order (A 6.4:286–87; Rauzy 1995, 40n26).

The connection between the simplicity of terms and the order of composition comes up in another passage: "A term which is anterior by nature is one which is obtained by substituting the simples for the composed. Or what is the same: naturally prior is produced by analysis; naturally posterior by synthesis" (C 241; Rauzy 1995, 34).[31] If we apply this to the case of fractions, it seems that Leibniz's point is this: the most simple are the unities, and since its fractions presuppose unities, they are more complex. This is due to the method of production, even if the operation is division rather than composition.

Does this approach regarding the fundamental role of arithmetical unity apply to the metaphysical unity of substances as well? In other words, is there an analogy here between the arithmetical unit and the metaphysical one? My contention is that, in the case of substances, a basic notion of unity is indeed (similarly) presupposed by any of its internal modifications. Let me try to exemplify this through Leibniz's notion of the law of the series.

As we know, for Leibniz, the law of the series defines a substance (see A 6.3:326 and section 1.4.2). Indeed, the law of the series has the right structure for this role; it expresses the conceptual priority of unity over the infinite multitude of states that it involves. In both mathematical and metaphysical contexts, the law of the series reconciles the tension between oneness and infinity discussed earlier (section 4.1). It thus shows how a multiplicity or an *infinity* of states can be reconciled with the fundamental unity of substance. This is articulated in terms of the dependency between the law of the series, seen as a law of generation, and each term in the series. In the realm of mathematics, such a law is a formula that determines an ordered sequence of terms in a series; in the metaphysical realm, it is seen as a single internal law that determines the unfolding states of a substance, following a set order. Thus, the law of the series makes a substance both unique and a single unity that remains the same through the change of its states.[32]

---

[31] "Terminus natura prior (posterior) est qui prodit pro composito (simplicibus) substituendo simplices (compositum). Sive quod idem est, natura prior prodit per analysin, natura posterior per synthesin: alter ex altero" (A 6.4:286).

[32] For the identification between the law of the series and the primary active force, see Leibniz's letter to De Volder from January 21, 1704, in which he writes, "[b]ut the persisting thing itself, insofar as it involves all cases, has primitive force, so that primitive force is like the law of a series, and derivative force is like a determination that designates some term in the series" (LDV 287). Let us note that a different understanding of the notion of form may also be consistent with its identification with the law of the series. For example, Arthur, who argues that forms do their work "by giving a teleological and functional unity over time, a diachronic rather than a synchronic unity" (R. T. W. Arthur, "Presupposition, Aggregation, and Leibniz's Argument for a Plurality of Substances," *Leibniz Review* 21 [2011]: 92), provides another dimension to

Among many other texts, in his correspondence with De Volder, Leibniz expounds the way a unity may admit a variety of states through the notion of the law of the series. Leibniz uses the law of the series in the correspondence as a principle of internal change, instructing the change in the substance's states, or rendering the transition from one perception to another.[33] Leibniz writes,

> The substance that succeeds is taken to be the same as long as the same law of the series, i.e., of the continual simple transition, persists that gives rise to our belief in the same subject of change, i.e., the monad. I say that the fact that there is a certain persisting law, which involves the future states of that which we conceive as the same, is the very thing that constitutes the same substance. (LDV 291)

The same line of reasoning appears in a letter from 1698 in which Leibniz addresses a question posed to him by Bayle, who asks how it is possible to compare simple and indivisible souls with a watch. Bayle argues that this comparison must be wrong since, if a simple being is composed of many pieces, like a machine, it would act diversely and the particular activity of each piece could at any moment change the course of all the others. In other words, what Bayle is asking is how the soul, being simple and indivisible, can contain a variety of states but at the same time does not amount to a composite of them. To clarify the sense in which a substance is not simply the sum of its states and cannot break down into them, Leibniz uses the law of the series:

> When it is said that a simple being will always do the same thing, a certain distinction must be made: if "doing the same thing" means perpetually following the same law of order or of continuation, as in the case of certain series or sequence of numbers, I admit that all simple beings, and even all composite beings, do the same thing; but if "same" means acting in the same way, I don't agree at all. Here is an example which explains the difference between these two senses: a parabolic movement is uniform in the first sense,

---

Leibniz's identification of the law of the series with substantial form. In particular, Arthur's understanding of the idea of form—as that which relates to the dynamic process of unfolding of states over time—can explain Leibniz's utilization of the idea of the law of the series in his letters to De Volder and its added value over the idea of form. For this point also see Phemister (2005, 122). In his forthcoming book *Ariadnean Threads*, Arthur extends this analysis to the relation between a monad and its perceptions.

[33] Leibniz writes to De Volder that substances are not "wholes that contain parts formally, but total things that contain partial things eminently" (January 21, 1704; LDV 289). See also GM III/2:500.

but not in the second, for the segments of a parabola are not the same as each other, as are those of a straight line. (WFN 83)[34]

As in the case of a basic mathematical unit, Leibniz expresses the relation between a unity and its diverse states through the idea of the law of the series. This is why the law of the series is often thought of as doing a *unifying* work.[35] However, I do not see the law as doing a *unifying* job in the sense of combining *separate* elements. Rather, the law of the series expresses substantial unity in the sense that it generates (and thus holds up) the internal complexity of the substance and thus preserves its unity. In this context, the oneness of substance means primarily *unity*. As Leibniz writes to Des Bosses, "the operation proper to the soul is perception, and the nexus of perceptions, according to which subsequent perceptions are derived from previous ones, forms the unity of the perceiver" (LDB 129). In this way, the internal complexity of the substance presupposes the law that defines its essence and informs its development.

4.5 THE DUAL SENSE OF "THE PLURAL PRESUPPOSES THE SINGULAR"

As the number three presupposes three units, so an aggregate of substances presupposes singular substances. Leibniz famously claims that "the plural presuppose the singular." But if the units presupposed by numbers (1 and 1 and 1) are not seen as numbers, why should the basic metaphysical unities be taken in a numerical sense? Thus, I would suggest that there is an analogy between the role that basic unities play in mathematics and in metaphysics. As numbers presuppose a basic unit, which is not itself considered a number, so substances presuppose singular substances (which are not numerically one). Hence, if there are aggregates of substances, there are clearly some singular substances. Indeed, this is almost trivial.

There is, however, another (somewhat less trivial) analogy between the priority of unity over internal complexity. Just as ½ presupposes a whole unit, so complex structures and sequences may presuppose a single whole. Likewise, in the

---

[34] As in other places (e.g., Leibniz's letter to De Volder from January 21, 1704), Leibniz's references to the law of the series can be thought of as related to the substance's diachronic (rather than synchronic) unity. As he explains to De Volder "from individuals there follow temporal things . . . all individual things are successive" (LDV 289).

[35] For example, Fleming claims that the law of the series "unifies the series of perception by determining or explaining the progress of the series" (N. Fleming, "On Leibniz on Subject and Substance," in *Gottfried Wilhelm Leibniz: Critical Assessment*, ed. R. S. Woolhouse [London and New York: Routledge, 1994], 2:118). Loemker notes that the law is that which shows that the whole is not "merely the aggregate of its parts" but "a unity which possesses reality beyond these changing modes themselves" (L 541).

metaphysical case, Leibniz regards a nested individual, involving infinitely many individuals, as a single unit, or individual (see chapter 6 for more detail on this). In this sense, the internal complexity (and/or multiplicity) of a substance presupposes the unity of the whole. For example, the qualities of a person presuppose the unity of the ego or the self that is the possessor of these qualities.

In aggregates, we typically find the picture of aggregated or collected pluralities (as number may be seen as a composition of units); but in dealing with substances, which are exemplified by the I or the ego or the complete animal,[36] it seems that another analogy between arithmetic and metaphysics might be at work—that is, that a single unity is presupposed by any of its fractions. A Leibnizian substance has infinitely many states; but perhaps Leibniz wants to emphasize that the multiplicity of states presupposes a single subject in which they all inhere, or a single unity whose states they are. This is nicely exemplified through Leibniz's use of the law of the series, so that each of the states presupposes the law generating all of them. The point is also supported by Leibniz's recurrent references to the ego as a true unit and being. The whole person, the ego, often comes up as Leibniz's prime example of a substance. And it always comes up as something that has a multiplicity of states, qualities, perceptions, and the like.

I suggest that Leibniz's dictum that the plural presupposes the singular (letter to Arnauld, GP II:97) should be read differently in each of these contexts. In the first case (which the literature has taken to be the *only* way of reading it), an aggregate presupposes a plurality of units; in the second case, a plurality of states presupposes a single unity in which all these states inhere. The latter seems to fit better with Leibniz's example of the ego, which is clearly not seen as a mere *collection* but, rather, as a single being whose unity is presupposed despite its internal complexity and the multiplicity of its states and constituents. And yet, on the standard reading, this example does not seem to fit with what it is supposed to exemplify—that is, the unity of the whole rather than the inference to a plurality of singular constituents from the existence of an aggregate. In other words, the aggregate argument, as it is often called, suits its name very well—that is, it applies to aggregates, herds, flocks, and the like, but as I would like to stress, it applies *only* to aggregates and not to substances such as human beings (which are, after all, one of Leibniz's prime example for a substance). One can infer, as Leibniz does, from the existence of aggregates to the necessity of substances composing them, but not from a substance (the ego or self, for example) to its constituents, because it is not composed of parts as an aggregate would be; rather, a substance has a natural unity that may still change its states. In this case, the qualities and states of a substance presuppose a whole, just as fractions presuppose a

---

[36] See, for instance, Leibniz's letter to Johann Bernoulli, September 20/30, 1698; AG 168.

unit, or as the states of a series presuppose its rule of generation. And in that sense, too, the multiple (states, qualities, etc.) presuppose the singular (substance) in which they all inhere.

A related suggestion (to be taken up in chapter 8) is that this conception of unity operates at different levels—that is, both at the level of the basic constituents of composed things (such as bodies and aggregates) and as the unity of complete beings such as humans; both levels require per se (or constitutive) unity, so that multiplicity, both internal and external, presupposes unity and/or unities. Following Fichant, I will suggest (in section 8.3) that it is this line of reasoning—that unity operates at both levels, at the bottom and at the top—that leads to the universalization of monads, so that we have monads both at the bottom level and at the top (and that some are more dominant than others).[37]

---

[37] The dominant monads are at the top and the less so are below them, but as this hierarchical structure goes *ad infinitum*, strictly speaking, there is no bottom.

5

INFINITY AND LIFE
A SKETCH OF LEIBNIZ'S DEVELOPMENT

5.1 INTRODUCTION

The notion of a natural machine, which remains a machine to the least of its parts, comes to the foreground as Leibniz's prime model of living beings in the "New System" in 1695.[1] This notion, as distinct from an artificial machine, will occupy the center stage of the next chapter. In the present chapter, I aim to trace some of the major steps Leibniz takes before coming to this intriguing way of drawing the line between living and nonliving things. I will first present a brief sketch of the way in which Leibniz uses infinity initially to describe, and ultimately to define, living beings.[2] In so doing, I will trace the development of the way Leibniz is using two concepts—infinity and life—that seem initially disparate until they come together in the way in which Leibniz draws the distinction between natural and artificial machines.

As will also become clear in this brief survey, according to Leibniz, to be living and active turns out to be a prerequisite for being a real entity. In other words, Leibniz comes to associate being with being animate, or activated by some soul-like thing—*anima*, entelechy, or substantial form, as he variously terms the source of activity and life in living beings. The relation between the notions of activity, force, and life will be further discussed in chapter 8. As argued in the previous chapter, for Leibniz, unity is also a necessary condition for being. What follows, then, is a brief sketch of how unity, infinity, and life come to be connected in Leibniz's notion of a true being or substance.

---

[1] On this issue, see M. Fichant, "Leibniz et les machines de la nature," *Studia leibnitiana* 35 (2003): 1–28; Duchesneau (2010); O. Nachtomy, "Leibniz on Artificial and Natural Machines," in *Machines of Nature and Corporeal Substances in Leibniz*, ed. Justin E. H. Smith and Ohad Nachtomy, The New Synthese Historical Library (Dordrecht: Springer, 2010), 61–80; Smith (2011); and Arthur (2014).

[2] For an extensive discussion of Leibniz's view of living beings, see Smith (2011).

5.2 EARLY THOUGHTS ABOUT INFINITY AND INDIVISIBILITY

As already noted in the first chapter, in his "Theory of Abstract Motion," Leibniz states some of his most fundamental presuppositions. Most of these presuppositions continue to inform the development of his later views. While these principles undergo some significant variations, they remain a constant feature of his understanding of the relations between infinity and unity. Most important for our concerns here are the following two principles:

(1) "There are actual parts in the continuum . . . and they are actually infinite."
(4) "There are indivisible or unextended beings" (A 6.2:264; L 139–40).

In his introduction to Leibniz's writings on the *Labyrinth of the Continuum*, Richard Arthur argues that these themes in Leibniz's early philosophy persist in his later views (while undergoing some transformations). He glosses them as follows:

1. The actual division to infinity and the postulation of indivisibles.[3]
2. Indivisibles are of unassignable quantity; they have quantity but it is unassignable, which allows for comparing them.
3. There are no minima or minimal parts.[4]

Some of these commitments, especially regarding the postulation of indivisibles, are related to Leibniz's mathematical work, and especially to his drawing on Cavalieri's method of indivisibles. Leibniz's metaphysical motivation for holding these commitments (especially 1 and 3) is related to an ancient concern—a concern that is clearly articulated in a note from the *De Summa Rerum* (of March 18, 1676), as follows:

> Whatever is divisible, whatever is divided, is altered —or rather, is destroyed. Matter is divisible, therefore it is destructible, for whatever is divided is destroyed. Whatever is divided into minima is annihilated; but that is impossible (A 6.3:392; DSR 45)

A persistent requirement in Leibniz's theory of true beings (or substances) is crisply articulated here: if something is to have persistent reality, it has to have true unity

---

[3] See also Garber (2009, 27–28).
[4] These are spelled out in LLC xxxii–xxxv as what Arthur calls Leibniz's first solution to the continuum problem.

or to be indivisible; for whatever is divided is destroyed. Leibniz thus links indivisibility with being, so that indivisibility is seen as a necessary condition for being. This supposition is also related to Leibniz's view, articulated in terms of the contrast between the intrinsic unity of a substance and the intrinsic divisibility (and hence multiplicity) of matter. Since matter is divisible (and thus plural), it cannot constitute the proper grounds for reality and cannot be considered a substance or a true being on its own. This is one of the central arguments Leibniz uses against Descartes's view that *res extensa* can be regarded as substances, and that extension is the fundamental attribute of bodies. As Leibniz will argue, extension is not fundamental; rather, it has to be derived from the nature of indivisible substances. Hence, both (secondary) matter and extension are not true beings but, rather, derive from the existence of true beings.[5] In this sense, a being must be an indivisible unity, which would also render it naturally indestructible; for if divided, its unity would be disrupted.

Along with these noted commitments, as early as his "Theory of Concrete Motion" (1670–71), Leibniz articulates the doctrine that, in every bit of matter, there are worlds within worlds, and that this goes on to infinity. In this context, the doctrine appears as a consequence of the infinite divisibility of the continuum. It is thus rather clear that, since his early writings, Leibniz is preoccupied with reconciling unity (or indivisibility) with multiplicity. As he writes,

> [f]or a unity always remains as great as it can [*unitas semper manet quanta maxima potest*], without destroying multiplicity [*salva multitudine*] and this happens if bodies are understood to be folded rather than divided. As, for example, a chord is one vibration [*ut chorda tremens una est*], even though there is no part of it that does not have its own particular motion. ("Metaphysical Definitions and Reflections," 1678–1680; A 6.4:1401; LLC 251)

Instead of the division of matter into bits or parts, Leibniz suggests here an ingenious model of folds.[6] He seeks to show that unity and multiplicity may be compatible. He is also using the powerful image of a vibrating string to illustrate that, while each of the string's parts are moving (and thus undergo some variation), the string remains one and the same. "Whoever understands this proposition well enough will laugh at the vain questions concerning the seat of soul." And just a few lines later, Leibniz adds, "[a] unity must always be added to multiplicity to the extent that it may" (A 6.4:1401;

---

[5] See, for example, GP IV:393–94; AG 250–51.
[6] See also Leibniz, *Definitiones cogitationesque metaphysicae*, 1678–1681 (?), A 6.4:1401, translation in Arthur (2001; LLC 249–51); Leibniz, *Conspectus libelli elementorum physicae*, 1678–79 (?), A 6.4:1900.

LLC 251). It is no accident, of course, that Leibniz's very definition of harmony, early and late, concerns the relations between diversity and identity (or the relations between multiplicity and unity). In his "Confession of a Philosopher" (1672–73), he writes that "harmony is unity in multiplicity" (A 6.3:122 and 146; *Confessio* 43–45 and 103), and in his letter to Wolff (May 18, 1715), harmony is seen as "the agreement between identity and variety" (AG 233–34). It is also worth recalling here Leibniz's famous comment in a letter to Sophie that his most fundamental meditations concern infinity and unity (GP VII:542).

Indeed, it is important to observe that Leibniz's early notion of an atom—the most fundamental and indivisible unit—is likened to a world. As he writes, "any atom will be of infinite species, like a sort of world, and there will be worlds within worlds to infinity" (A 6.2 N40; LLC 338). In contrast to contemporary intuitions, and to what may be suggested by the literal sense of the word, the notion of atom that Leibniz is supposing here implies that any atom involves internal richness and multiplicity. This is consistent with the view presented in the previous chapter—that is, that internal richness may presuppose a single unit, in analogy to the way fractions presuppose a most basic unit whose fractions they are. But this sense of atom is not at all peculiar to Leibniz; rather, it appears frequently in the period, as can be seen in Malebranche and Pascal, for example. Leibniz articulates this view again several years later in his notes from Paris (1676). Here, we find him writing that every part of the world, regardless of how small, "contains an infinity of creatures," which is itself a kind of "world" (A 6.3:474).[7] In another curious and revealing passage from the same period Leibniz writes,

> [w]e must try to see if it can be demonstrated that there is something infinitely small, yet not divisible. If such a thing exists, there follow some wonderful consequences about the infinite: namely, if we imagine creatures of another world that is infinitely small, we will be infinite in comparison with them. Whence it is clear in turn that we could be imagined as being infinitely small in comparison with another world that is of infinite magnitude, and yet bounded. Whence it is clear that the infinite is—as of course, we commonly take for granted—something other than the unbounded. This infinite should more properly called *Immensum*. ("On the Secrets of the Sublime," A 6.3:475; LLC 49–51)

As we have seen in chapter 3, on some occasions Leibniz uses the term *Immensum* to indicate a kind of infinity that is literally beyond measure—that is, that which

---

[7] DSR 25.

cannot be measured because it does not belong to the category of quantity (and not because it is very large as current English usage of "immense" suggests). Here we see that Leibniz is clearly looking for something that may be infinitely small yet indivisible.

In the dialogue "Pacidius to Philalethes," written on board Prince Rupert's yacht in November 1676 on his way back from Paris to Germany via the Netherlands, Leibniz continues to express the view we have seen earlier: "in any grain of sand whatever, there is not just a world, but an infinity of worlds" (A 6.3:566; LLC 211). Leibniz might well have been aware of the earlier microscope observations by Hooke (published in 1665) and of examples such as Pascal's *ciron* (see chapter 7). But in Holland he would find some empirical support for the "worlds within worlds" thesis in the work of Swammerdam and Leeuwenhoek.[8] Thus we can see that Leibniz was not only endorsing the view that there are worlds within worlds very early in his career; he was also relating this view to the discoveries of *animalcula* made by the early microscope observers. Indeed, he often refers to these observations as confirming his conviction that living beings are to be found everywhere in nature. Given his search for infinitely small but indivisible unity, it is not difficult to understand Leibniz's enthusiasm about the newly discovered world of creatures, revealed under the lens of the microscope. At the same time, it is also important not to overstate this point. Leibniz must be well aware that observing smaller and smaller *animalcula* does not support the claim that this would go to infinity.[9] The division of organic bodies to

---

[8] See Arthur (2014, 69).

[9] See Smith (2011, 155); and O. Nachtomy, "Infinity and Life: The Role of Infinity in Leibniz's Theory of Living Beings," in *The Life Sciences in Early Modern Philosophy*, ed. O. Nachtomy and J. E. H. Smith (New York: Oxford University Press, 2014), 19. There is some textual evidence that Leibniz himself was well aware of the fallacy of arriving at the infinite on the basis of experience. For example, in a letter to Rémond (November 4, 1715), Leibniz writes: "The author [the reference is to Du Tertter's *Réfutation* of Malbranche's system published in Paris in 1715] adds (vol. 1, 307) that in the so-called knowledge of infinity, the mind only sees that lengths can be placed end to end and be repeated as much as one would like. Very well, but this author might consider that knowing that this repetition can always be made already amounts to knowing infinity" (GP III:658–89; English translation is from http://www.leibniz-translations.com/remond1715.htm). Leibniz's point here is that the idea that the infinite is based on thinking that one could go on indefinitely, presupposes (rather than yields) the notion of infinity. While this is written in a different context, it is clear that the same argument would apply to observation of smaller and smaller animals, and it supports the point that Leibniz should not base his claim that the structure of living beings is infinite on microscopic observations alone. More direct evidence can be found in the following passage: "Although the conservation of the animal is favored by the microscopes, nonetheless we were aware of small bodies before their discovery, and thus we were already very well able to foresee the small animals, as Democritus foresaw the imperceptible stars in the Milky Way before the discovery of the telescope" (Leibniz 1985, V/2:302; translation in Smith 2011, 222).

infinity, the existence of microscopic animals, and Leibniz's concern that, owing to the division to infinity there would be no elements, are expressed again in Leibniz's letter to Malebranche of 1679:

> There is even room to fear that there are no elements at all, everything being effectively divided to infinity in organic bodies. For if these microscopic animals are in turn composed of animals or plants or other heterogeneous bodies, and so on to infinity, it is apparent, that there would not be any elements. (A 1.2:719, translated by Smith 2011, 160)

Leibniz's deeper commitment is that there are such elements, and his task is to provide an account of how this might work. At the same time, Leibniz's reference to indivisible units (or atoms) remains quite ambiguous in his early writings. Are these atoms supposed to be physical, unsplittable bits of matter, or are they supposed to be more akin to geometrical points, or to minds, as he implies in some texts? Since all these notions appear in Leibniz's early writings, it is not so easy to decide this question, especially since Leibniz's reasoning is often related to the infinite divisibility of matter.

5.3 THE MIDDLE YEARS: UNITY AND ANIMATION

But precisely because of the central role this argument plays, Leibniz would soon come to see that grounding any notion of indivisibility in material atoms is doomed, if the very nature of matter is to be divisible. In the "New System" (as well as later in "On Nature Itself," for example), Leibniz will make it very explicit that his atoms are substantial (rather than material or merely mathematical). These would become atoms of substance whose unity is related to their inner force and inherent activity. As Daniel Garber has argued recently, it is very plausible that Leibniz developed this insight through his reading of Cordemoy in 1685.[10] But this idea comes up earlier, as well. For example, in the *Conspectus libelli elementorum physicae* of 1678–79 (A 6.4:1986 N365), we already find a clear statement relating the basic requirement for unity to a soul-like thing and to being animated. Leibniz writes,

> [n]ow there follows the subject of incorporeals. There turn out to be certain things in body which cannot be explained by the necessity of matter alone. Such are the laws of motion, which depend on the metaphysical principle of cause and effect. Here therefore the soul must be treated, and it must be

---

[10] Garber (2009, 68–70).

shown that all things are animated (*omnia esse animata*). Unless there were a soul, i.e. a kind of form, a body would not be an entity, since no part of it can be assigned which would not again consist of further parts, and so nothing could be assigned in a body which could be called *this something, or some one thing* [*hoc aliquid, sive unum quidam*].[11] (A 6.4:1988; LLC 233–35)

In a text that the Academie editors date to 1683–1685, Leibniz writes, "unless it is animated, or contains within it a certain single substance, corresponding to the soul, which they call substantial form, or primary entelechy, body is no more one substance than a woodpile" (A 6.4:559; LLC 265). The woodpile is easily recognized as one of Leibniz's paradigmatic and recurrent examples of an aggregate—that is, a collection of things that lacks true unity. Here, as elsewhere in this period (A 6.4:1464; LLC 257–59), Leibniz continues to worry that, if a body is divisible and actually divided, it would not be a real entity but would be a mere phenomenon. What comes into the foreground in this period is the conviction that such indivisible unities are also animated. At the same time, we should not lose sight of the fact that being animated (*animata*) literally means to be activated by an *anima*—that is, a soul-like principle or the principle of life (in living things), reminiscent of Aristotle's account. It is also no accident that this terminology appears in Leibniz's polemics with the Cartesians, who attempt to explain matter purely geometrically, in terms of extension alone, such that everything in nature could be explained without invoking the "obscure forms" of the Aristotelian/scholastic tradition. Thus, commenting on the writings of Cordemoy, whom he tellingly calls a "semi-Gassendist," Leibniz makes the following remark:[12]

> if all organic bodies are animated [*omnia corpora organica sunt animata*], and all bodies are either organic or collections of organic bodies, then it follows that all bulk is divisible, but that substance itself can neither be divided nor can it be destroyed. (1685; A 6.4:1798; LCC 277)

Here, too, it seems that intrinsic unity can come only from an animate, soul-like thing. And again it is the substance's indivisibility that makes it indestructible.

---

[11] Incidentally, in this text we also find Leibniz likening *this some one thing*, the mind, to a perceiving mirror—an image that we shall have the opportunity to explore in chapter 7. As he writes, "[t]here are as many universal mirrors as minds, for every mind perceives the whole universe, but confusedly" (1678–79; LLC 235).

[12] "It is noteworthy that not only the common Cartesian, who call everything that is extended divisible but even the semi Gassendist, who deems every substance indivisible and truly one, appeal to ideas" (LLC 277).

Leibniz's original position here consists in moving from a common (and, more precisely, Aristotelian) notion of animation, one that applies to organized living things which are self-nourishing and self-moving, to a notion of animation that indicates the work of a nonphysical, indivisible, and active soul-like thing.[13] This Leibnizian notion of animation is more adequately seen as metaphysical rather than as biological. Leibniz's application of this notion of animation is certainly much more universal than the limited scope of the principle of life in Aristotle, in that it is not limited to accounting for the traditional phenomena associated with living beings.

In "Primary Truths" (1689–90), we find once again Leibniz's view that, "every particle of the universe contains a world of an infinity of creatures" (A 6.4:1648; AG 34).[14] But at the end of this piece, we find Leibniz's idea (which will reoccur in the "New System") that "animate things neither arise nor perish but are only transformed." This indicates—as will become transparent in later writings—that Leibniz is working with a notion of living (or animate) that is very different from both ours and that of his contemporaries—it is not a notion of living things that naturally come to life and are bound to die; rather, Leibniz's notion of life is linked to the divine creation of active individuals, so that, once created, living creatures cannot naturally perish (but can only be annihilated by God).[15] It is a notion of living beings, whose existence corresponds to that of the world. Or, to put it differently, the existence of the world depends on the existence of animate beings.

In the "Discourse on Metaphysics," article 32, we find Leibniz's idea that the soul (*l'âme*) constitutes "a perpetual living expression" of the world. This is of course related to Leibniz's thesis (stated explicitly in article 9 of the "Discourse on Metaphysics") that each substance is "like a complete world and like a mirror of God or of the whole universe, which each one expresses in its own way ... and that every substance bears in some way the character of God's infinite wisdom and omnipotence and imitates him as much as it is capable" (AG 42; A 6.4:1542). The connection between living and being an expression of God and the universe, however, is often overlooked. Indeed, it is not until some ten years later that Leibniz would pen the expression "living mirror" (*miroir vivant*) in his comment on Pascal. I will explore this connection at length in relation to Leibniz's notion of the living mirror in chapter 7.

---

[13] For an elaboration of this point, see R. Andrault, *La vie selon la raison. Physiologie et métaphysique chez Spinoza et Leibniz* (Paris: Champion, 2014), sec. 3.1.4.

[14] In a note on a letter of Michelangelo Fardella from 1690, we find again Leibniz's commitment to the worlds within worlds *ad infinitum*. He writes: "there are substances everywhere in matter, just as points are everywhere in a line.... [J]ust as there is no portion of a line in which there is not an infinite number of points, there is no portion of matter which does not contain an infinite number of substances" (AG 105).

[15] See also "Discourse on Metaphysics," art. 9 (AG 42) and art. 32 (AG 64).

But let me observe here that the very notion of living, i.e. power of action, may well be one of the aspects in which a created substance is seen to imitate God's omnipotence (in a limited way, of course).

In his correspondence with Arnauld and related texts, it becomes clear that, when he refers to the creatures that are to be found everywhere to infinity, Leibniz means living, active things; it is also clear that there is something about their being alive that is responsible for their ability to provide true unity and (so to speak) resist division. As he writes,

> and since matter is infinitely divisible, no portion can be designated so small that it does not contain animated bodies, or at least bodies endowed with a primitive Entelechy or (if you permit me to use the concept of life so generally), with a vital principle; in short, corporeal substances, of all of which one can say in general that they are living. (GP II:118)[16]

In a letter to Arnauld from November 28, 1686, Leibniz goes on to suggest that living (or being animate) is the mark of a true corporeal substance: "I cannot say precisely whether there are true corporeal substances other than those that are animated, but souls at least serve to give us some knowledge of others by analogy" (AG 79; GP II:76). Leibniz holds here that animate things are the only corporeal substances that we know. In the same letter, he clarifies that only animate beings can be the source of true unity in light of the actual division of matter *ad infinitum*:

> For the continuum is not merely divided to infinity, but every part of matter is actually divided into other parts as different among themselves as the aforementioned diamonds. And since we can always go on in this way, we could never reach anything about which we could say, here is truly a being, unless we found *animated machines* whose soul or substantial form produced a substantial unity independent of the external union arising from contact. (AG 80; GP II:77, emphasis added)

The only things that can be regarded as true beings are animate machines whose soul or substantial form produces substantial unity rather than mere contiguity. As a consequence, they are the only limit, so to speak, to the infinite divisibility of matter.

---

[16] But see A II/2:249, which does not contain the word "entelechy": "où il n'y ait dedans des corps animés, ou au moins informés, c'est à dire des substances corporelles." For an explanation of this addition see George Le Roy edition (LR), 311n24. I thank Raphaele Andrault for drawing my attention to this point.

While organic bodies, taken apart, are divisible, animate things are indivisible by virtue of the substantial unity produced by their activity. Unlike Pascal, for example, who claims that infinite divisibility lurks in living things as well (see chapter 7), Leibniz considers living beings to be the ultimate unities—unities that are indivisible and indestructible precisely because they are animate and active.[17] Of course, the bodies of living things are actually divided, but as units of substance, living beings are not divided; in fact, they are indivisible. These animate unities are the grounds of whatever reality there is in nonliving things and remain at the foundation (both ontological and explanatory) of any visible transformation and variation in nature:

> Those who conceive that there is almost an infinite number of little animals in the smallest drop of water, as the experiments of M. Leeuwenhoek have made known, and who do not find it strange that matter is everywhere full of animate substances, will not find it strange that there is something animate even in ashes and that fire can transform an animal and reduce it in size instead of totally destroying it . . . [such animals are] little organic bodies, wrapped up as they are because of a sort of contraction from a larger body which has undergone corruption. (Leibniz to Arnauld, October 9, 1687; GP II:122)

As these passages show, in his correspondence with Arnauld, Leibniz is using the notion of life (or, more precisely, being animate) as a mark of being a true substantial unity. The notion of life is now connected with that of infinity, such that living beings are to be found everywhere (or "in the least part of matter")[18] and such that they form true unities that are not susceptible to the infinite divisibility of matter. In the "New System" (1695), the relation between infinity and life becomes much more explicit. There, Leibniz is no longer using infinity merely to describe nature as worlds within worlds to infinity; rather, infinity becomes one of the defining features of living beings. In the "New System," Leibniz articulates the distinction between living and nonliving things in terms of the distinction between natural and artificial machines. This distinction is the topic of the next chapter.

---

[17] As Raphaele Andrault has recently put this: "à partir des lettres à Arnauld, ce n'est plus l'unité véritable des corps qui constitue le critère d'attribution des formes substantielles, mais, presque au contraire, une organisation corporelle composée, divisible mais cohérente, en tant qu'elle indique une unité non corporelle, qui doit se penser par analogie avec l'esprit: celle des animaux, et par extension, de tous les 'vivants.' Ainsi, alors que les corps eux-mêmes sont des 'cadavres,' c'est-à-dire des agrégats ou des machines, ils forment, avec l'âme, des 'vivants,' des 'animaux' ou des 'hommes' qui ont une vraie unité" (Andrault 2014, 113).
[18] GP II:111–29.

5.4 COMPLETE CONCEPTS, LAW OF THE SERIES, AND INTERNAL POWER OF ACTION

Before we turn to examine this distinction in detail, I would like to consider another aspect of the infinity of individual substances. Since the early 1680s or so, and stemming from his work on logic and language, as well as the nature of possibility, predication, and truth, Leibniz defines individual substances through their complete concepts—concepts that involve all the individual's predicates—past, present and future. Leibniz clearly regards such concepts as infinite. As he famously states in "Primary Truths," "Every individual substance contains in its perfect notion the entire universe and everything that exists in it, past, present, and future" (AG 32). In the "Discourse on Metaphysics," on the basis of the *in-esse* principle, according to which every true proposition is such that the predicate term is included in the subject term, Leibniz argues that,

> the nature of an individual substance or of a complete being is to have a notion so complete that it is sufficient to contain and to allow us to deduce from it all the predicates of the subject to which it is attributed. . . . Thus when we consider carefully the connection of things, we can say that from all time in Alexander's soul there are vestiges of everything that has happened to him and marks of everything that will happen to him and even traces of everything that happens in the universe, even though God alone could recognize them all. ("Discourse on Metaphysics," art. 8; AG 41; A 6.4:1541)

And again, in article 13:

> the notion of an individual substance includes once and for all everything that will happen to it and that, by considering this notion, one can see there everything that can be truly said of it, just as we can see in the nature of a circle all the properties that can be deduced from it. ("Discourse on Metaphysics," art. 13; AG 44; A 6.4:1546)

Leibniz's view that, in each individual, "there are vestiges of everything that has happened to him and marks of everything that will happen" is related to his doctrine of the law of the series. As Leibniz describes this connection early on, "[t]he essence of substances consists in the primitive force of acting, or in the law of successive changes, as the nature of a series in numbers" (A 6.3:326).[19] This doctrine has a long

---

[19] "L'essence des substances consiste dans la force primitive d'agir, ou dans la loy de la suite des changemens, comme la nature de la series dans les nombres" (comments on Foucher regarding Malebranche; A 6.3:326).

history in Leibniz's thought. It goes back to his very early work on the principle of the individual from 1663, in which he concludes that the adequate source of individuation is the complete or total entity. The doctrine of marks and traces is also clearly expressed in a note entitled "Meditation on the Principle of the Individual" composed in 1676 (A 6.3:490–91; DSR 51), where Leibniz considers two perfectly similar squares and argues that the only way to distinguish between them is through a different means of production. For example, such seemingly identical squares can be produced by joining two rectangles or by joining two triangles. Thus, not only the current appearance or state of a thing has to be considered but its whole sequence of generation. It follows, he writes, that "there is present in any matter something which retains the effect of what precedes it, namely, a mind." And, therefore, "right up to the present, there is in it a quality of such kind as to bring it about" (A 6.3:491).

In previous work (2007a, sec. 2.4), I have argued that a complete concept of an individual may be seen as a unique program for the development of an individual substance. A complete concept provides a conceptual blueprint for actual things, that is, a complete concept prescribes a possible course of action and thus nicely captures Leibniz's notion of a possible individual—which he regards as a candidate for actualization. As we know, for Leibniz, sets of possible individuals make up possible worlds, and not all possible worlds are realized. Thus, complete concepts would also account for Leibniz's view that God preconceives all possible worlds in advance of creation, and would opt to realize the most harmonious of them (which is the best of all possible worlds).

It is important to note that the Leibnizian notion of an individual involves two components: on the one hand, a program of action, defined through its complete concept, and, on the other, a source of action, an entelechy, that realizes the course of action in bringing about the change of states in an individual. Thus, Leibniz's definition of individuals in terms of their complete concepts does not conflict with his definition of an actual individual substance in terms of its internal source of activity—a first entelechy or force that enables it to realize its unique course of action. Rather, upon creation, the law of the series that generates a possible course of action in God's mind would become active and serve as the individual program of action; it would thus realize the same course of action in a created, individual substance, and would in fact render a possible course of action actual. In this sense, the production rule of an individual may be considered as the forerunner of the source of unity and activity in created substances—the same source that the tradition which Leibniz attempts to rehabilitate (as he makes clear in the "New System") assigned to the substantial form.[20]

---

[20] On the connection between the law of the series and the substantial form, see Phemister (2005, 195, 122). Phemister identifies the law of the series with the substantial form.

It is also important to see that the notion of activity and life—a first entelechy or a source of activity—does not exclude or replace the logical definition of an individual substance in terms of its complete concept; rather, these definitions may well exist side by side and are certainly compatible. Indeed, I would argue that they are complementary. Thus I fully agree with Maria Rosa Antognazza, who writes that,

> If . . . the sustained study of physics and the development of the *Dynamica* had led Leibniz to present his theory of substance in a new light [in the "New System"], . . . his account seems to complement, rather than to compete with, the "complete concept" theory of substance more directly related to his logical investigations.[21]

While the complete concept serves to define a possible individual, the notion of activity pertains to actual substances. Indeed, primitive force or power of action is precisely what's needed to render a possible individual actual (see Nachtomy 2007a, ch. 5). But while the definition of an individual as an infinite sequence of predicates remains in the background, the notion of life (or primitive force of action) as the other essential component to the definition of actual individual substances, comes to the foreground.[22] This is of course closely related to Leibniz's rehabilitation of substantial forms and his use of force as a defining feature of substance—a project he has already announced by 1679, but mainly executed through a series of articles published in 1695–1698.[23]

Thus, both definitions of the individual, that by means of a complete concept and that by means of internal force (or power of action), complement one another. At the same time, their mutual function is made intelligible by virtue of the law of the series. As noted, on my reading, the law of series is operative both in the context of generating the unique sequence of predicates in God's mind (thus making up an individual concept) and in the context of informing the development of an individual

---

[21] Antognazza (2009, 349).

[22] Indeed, the terminology of the complete notion is no longer present in Leibniz's later writings. And, as we have just seen, it is a notion of activity and animation that seems to be responsible for unity. I agree with Fichant that, in the correspondence with Arnauld, Leibniz's notion of "complete being" (art. 8 of the "Discourse on Metaphysics") is reformulated in terms of a complete indivisible being (*être accompli indivisible*). Fichant further argues that, in this period, the indivisibility of a substantial form becomes the main manifestation of substantial unity, which has been expressed by the logical unity of the [complete] notion in the "Discourse on Metaphysics" (Fichant 2004, 87).

[23] The project involves Leibniz's critique of the Cartesian law of the conservation of the quantity of motion, replacing it with his claim for the conservation of living force, *vis viva*, expressed by mv^2. See A 1.2:225; and Garber (2009, 128–32); Antognazza (2009, 345–54).

substance, once it is created. In other words, the law of the series is essential both to the conception of a possible individual and to its realization.

As we shall see, the doctrine of the law of the series is also of crucial importance for Leibniz's definition of living beings in terms of natural machines (machines whose parts remain machines). Before getting into the details in the next chapter, here is a brief preview. In his recent book, *Leibniz*, Richard Arthur writes,

> what makes a natural machine "the *same* machine in its least parts" for Leibniz is its possession of a substantial form or monad. It does not have to have the same parts from one instant to another, so long as the parts it does have contribute to its own functions and end. For this it needs to be the source of its own actions, and also to have a law or "program" for the development and unfolding of these actions. Each of these two aspects of Leibnizian forms is crucial.[24]

As I have argued in previous work, and as Arthur nicely states, "it is the internal law governing the unfolding of the states of a substance that accounts for it having a genuine unity, as opposed to the accidental unity of an artificial machine."[25] What gives a natural machine—a machine with an infinitely complex structure—its unity is its internal law. This internal law functions as a program for self-organization and self-regulation in the sense of instructing its change of states, so that each Leibnizian substance remains a self-sufficient unit. According to Leibniz, then, a living being is infinite in the sense that it is always active, always developing and changing, and, particularly, with respect to its inner nested structure that develops *ad infinitum*. The infinity and unity of living beings is thus intrinsically related to their infinite creator, and consequently, their being "divine machines."[26] It is to a study of this notion that we now turn.

---

[24] Arthur (2014, 73).
[25] Ibid. See also Nachtomy (2007b).
[26] See also Leibniz's Fifth Letter to Clarke (arts. 115 and 116, in AG 344–45).

6

ANIMATE AND INANIMATE THINGS, NATURAL AND ARTIFICIAL MACHINES

6.1 INTRODUCTION

As we have seen in the previous chapter, from the "New System" in 1695, Leibniz spells out the distinction between living and nonliving things in terms of a subtle difference between natural and artificial machines.[1] In many of his subsequent writings, Leibniz describes living beings as machines, and each of its parts remains a machine to infinity.[2] I call this the nested structure of natural machines. As we shall see, according to Leibniz, the nested structure *ad infinitum* is the main intrinsic difference between a natural machine, which is God's creation, and an artificial machine, which is a work of human art. Leibniz holds that the distinction turns on this: unlike an artificial machine, a natural machine remains a machine in the least of its parts. My main objective in the present chapter is to cast some light on what Leibniz means by this phrase. The upshot is to highlight and clarify an important aspect of his view of living or animate beings that turns on his peculiar usage of infinity in this context.

Before attending to Leibniz's distinction in detail, I will present (in section 6.2) the immediate context for Leibniz's distinction—that is, the view to which Leibniz is responding. In the "New System," Leibniz is responding to Descartes's attempt to

---

[1] There are some precedents to Leibniz's notion of natural machine, such as in his letter to Bossuet of 1692 or even earlier in his notes on Cordemoy dating from 1685, where he argues that not even an angel could create a human being. But the use Leibniz makes of this distinction here, as drawing a line between animate and inanimate things, is new and radical. In the earlier texts, Leibniz is still operating within the more common picture of worlds within worlds, but in the "New System" and the *Monadologie*, the functional and structural relations to infinity are emphasized as typical of animate vs. inanimate things. Cf. Andrault (2014, sec. 3.2.1) and Fichant (2003).

[2] In a letter to Lady Masham from 1704, Leibniz writes, "I define an organism or a natural machine, as a machine each of whose parts is a machine, and consequently the subtlety of its artifice extends to infinity, nothing being so small as to be neglected, whereas the parts of our artificial machines are not machines. This is the essential difference between nature and art, which our moderns have not considered sufficiently" (GP III:356).

use the distinction between artificial and natural machines in his reductive program. Descartes wants to argue that the difference between artificial and natural machines is no more than a difference in degree of complexity and the subtlety of parts. In this picture, natural machines such as animals are simply very complex machines with minute parts that are invisible to us. In section 6.3, I present Leibniz's distinction between natural and artificial machines. In section 6.4, I question the coherence of his distinction. In section 6.5, I develop a structural reading of Leibniz's notion of a natural machine, and in section 6.6, I develop a functional reading of this notion. In conclusion, I will suggest that both readings (structural and functional) are compatible and that both illuminate Leibniz's definition of a natural machine as a machine remaining a machine in the least of its parts, and thus distinct from inanimate machines. This will set up the background for raising the question of which kind of infinity applies to natural machines and which to artificial machines. This question will be addressed more fully in the following chapters.

While Leibniz's distinction between artificial and natural machines has significant consequences for his metaphysics, it turns on a very subtle nuance. The first point to notice here is that Leibniz describes both natural and artificial things as machines— that is, in mechanistic terms. This seems consistent with the Cartesian program to describe the natural world in mechanistic terms. Descartes's program to describe animals (as well as the human body) as subtle and complex machines that need no vital forces, let alone minds, over and above the structure of their organs, is particularly relevant here.[3] By contrast, Leibniz's agenda may be seen as an attempt to revive the Aristotelian distinction between animate and inanimate things in "an intelligible way" and to resist the Cartesian reduction of natural machines to artificial ones.[4] It is with this aim in mind that Leibniz draws the distinction between artificial

---

[3] In his *Principles of Philosophy,* part 4, art. 203 (CSM I:288–89; AT VIIIA:326), Descartes seems to assimilate the artificial and the natural. For him, artificial machines serve as models to explain natural models. Natural machines are much more complicated, but he wants to establish that they are ultimately of the same kind as artificial machines. Descartes uses the notion of divinely created machines to show that the subtle parts of machines are extremely complex and invisible to us. While both Descartes and Leibniz argue that machines are extremely subtle, Descartes uses this point to argue for his view that, in the final analysis, animals are nothing but subtle machines. By contrast, Leibniz uses this point to argue that there is a categorical difference between them. See also Descartes's *Treatise on Man* in its entirety, and *The Passions of the Soul,* first part, arts. 5 and 6, where he writes, e.g., that the body has in it the corporeal principle of movement (art. 6; CSM I:329–30; AT XI:330–31).

[4] See, for example, Leibniz's controversy with Stahl: "And in this it seems that the ancients had already discerned something when they said that nature does nothing in vain, but tends toward an end, and other things of this sort that the moderns wrongly disapproved, as if the nature of bodies were nothing other than mechanism. They had little consideration for the fact that God the Author has directed all things toward ends, and that the actions of the soul

and natural machines in the "New System" of 1695—a work in which he attempts to reconcile the ancient and the modern philosophies of nature (in essence, by accepting mechanical description at the level of physics and Aristotelian description at the level of metaphysics). Thus, while Leibniz accepts a mechanical description of bodies, he strongly resists the Cartesian attempt to fully describe natural machines in terms of artificial ones. As he writes,

> I am the most readily disposed person to do justice to the moderns, yet I find that they have carried reform too far, among other things, by confusing natural things with artificial things, because they have lacked sufficiently grand ideas of the majesty of nature. (AG 141–42; GP IV:481)[5]

To better understand what Leibniz is resisting here, let us briefly review the radical reform suggested by Descartes. This will clarify why Leibniz thinks that it was carried too far, "by confusing natural things with artificial things."

6.2 DESCARTES AND THE ANALOGY BETWEEN NATURAL AND ARTIFICIAL MACHINES

Descartes's agenda in his projection of a new science was clear and ambitious. He sought nothing less than full mechanization of the natural world. More precisely, he sought mechanization of our view of the natural world in the sense that it should be described in terms of extended matter in motion. In effect, Descartes sought to replace any reference to incorporeal agencies, such as powers, faculties, or forms in the explanation of nature with the quantitative and measurable features of extended matter in motion. In this way, the natural world—at least the part belonging to *res extensa*—would be described in purely geometrical/quantitative terms.

One of the most difficult tasks facing Descartes's program was to provide an account of the phenomenon of life and especially of some features of living things such as nutrition, growth, and generation, which were traditionally explained by reference to a vegetative and sensitive soul. Descartes supposed that nature always acts in accordance with the laws of mechanics. He thus attempted to show that vital force is reducible to heat in the heart, understood as matter in motion. Likewise, he argued

---

are perfectly coordinated with corporeal actions while the soul makes use of perception and appetite" (LSC 27).

[5] See Leibniz's letter to Clarke (GP VII:377), where he mentions our tendency to ignore the grandeur of nature as stemming from an infinite author.

that any movement in the bodies of animals can be explained by attending to the mere disposition of their organs.

As Descartes writes in the preface to his *Description of the Human Body*,

> [i]t is true that we may find it hard to believe that the mere disposition of the bodily organs is sufficient to produce in us all the movements which are in no way determined by our thought. So I will now try to prove the point, and to give such a full account of the entire bodily machine that we will have no more reason to think that it is our soul which produces in it the movements which we know by experience are not controlled by our will than we have reason to think that there is a soul in a clock which makes it tell the time. (CSM I:315)[6]

As Gary Hatfield clarifies, "Descartes' aim was to mechanize virtually all of the functions that had traditionally been assigned to the vegetative and sensitive souls," and, "[t]o a large extent, Descartes physiology may be seen as a straightforward translation of selected portions of previous physiology into the mechanistic idiom."[7] Descartes's attempt to mechanize all the functions that had traditionally been assigned to the vegetative and sensitive souls comes out clearly in the conclusion to his *Treatise on Man*:

> these functions follow from the mere arrangement of the machine's organs every bit as naturally as the movements of a clock or other automaton follow from the arrangement of its counter-weights and wheels. In order to explain these functions, then, it is not necessary to conceive of this machine as having any vegetative or sensitive soul or other principle of movement and life, apart from its blood and its spirits, which are agitated by the heat of the fire burning continuously in its heart – a fire which has the same nature as all the fires that occur in inanimate bodies. (CSM I:108)[8]

---

[6] "Il est vrais qu'on peut avoir de la difficulté à croire que la seule disposition des organes soit suffisante pour produire en nous tous les mouvements qui ne se déterminent point par notre pensée; c'est pourquoi je tâcherai ici de le prouver, et d'expliquer tellement toute la machine de notre corps, que nous n'aurons pas plus de sujet de penser que c'est notre âme qui excite en lui les mouvements que nous n'expérimentons point être conduits par notre volonté, que nous en avons de juger qu'il y a une âme dans une horloge, qui fait qu'elle montre les heures" (AT XI:226).

[7] G. Hatfield, "Descartes' Physiology and its Relation to His Psychology," in *The Cambridge Companion to Descartes*, ed. John Cottingham (Cambridge: Cambridge University Press, 1992), 341–43.

[8] "[C]es fonctions suivent toutes naturellement, en cette Machine, de la seule disposition de ses organes, ne plus ne moins que font les mouvements d'un horloge, ou autre automate, de

While Descartes's aim is very clear and even somewhat simplistic—namely, to show that all the phenomena evident in living things can be explained mechanistically—his argumentative strategy is very sophisticated. The first step in his strategy is to conceive of all animals (as well as the human body) as *machines*. Once the body of an animal is referred to as a machine, Descartes appeals to the comparison between a machine manufactured by humans and a machine created by God. Roughly stated then, Descartes's strategy is to model natural machines on artificial ones. More precisely, he argues that the differences between the workings of a complex artificial machine, such as a clock or a fountain, and those of animal bodies are only apparent and turn on their degree of subtlety alone. Descartes attempts to show that, in essence, complex machines and animal bodies are of the same kind, and that the only differences between them reduce to degrees of complexity and the subtlety of their parts. Thus just as we don't need to invoke an occult agency in a clock that shows the hour, so there is no need to invoke such agency in our body other than the dispositions of its organs and parts. This is all the more true in animals and plants, which lack rational functions. Both functionality and the movement of animals can be ascribed to their internal workings, just as the operations of complex machines are.

As Descartes states in the *Principles of Philosophy*, part 4, article 203,

> I do not recognize any difference between artefacts and natural bodies, except that the operations of the artefacts are for the most part performed by mechanisms which are large enough to be easily perceivable by the senses—as indeed must be the case if they are to be capable of being manufactured by human beings. The effects produced in nature, by contrast, almost always depend on structures which are so minute that they completely elude our senses. Moreover, mechanics is a division or special case of physics, and all the explanations belonging the former also belong to the latter so it is no less natural for a clock constructed with this or that set of wheels to tell the time than it is for a tree which grew from this or that seed to produce the appropriate fruit. (CSM I:288)

It is mainly to this powerful and influential attempt to reduce natural machines to artificial machines that Leibniz responds. It is worth stressing that Leibniz does not

---

celle de ses contrepoids et de ses roues; en sorte qu'il ne faut point à leur occasion concevoir en elle aucune autre Ame végétative, ni sensitive, ni aucun autre principe de mouvement et de vie, que son sang et ses esprits, agités par la chaleur du feu qui brûle continuellement dans son cœur, et qui n'est point d'autre que tous les feux qui sont dans les corps inanimés" (AT XI:202; see also AT VI:45–46).

object to Descartes's seeing both artificial and natural as subtle machines. Rather, he attempts to draw a distinction between them as two distinct *types* of machines.

6.3. LEIBNIZ ON NATURAL AND ARTIFICIAL MACHINES

In the "New System," Leibniz insists that natural machines have something substantial—*soul or form*—that makes them one and the same thing in the *least of their parts*. Leibniz's formulation is that a natural machine, unlike an artificial one, "remains machine to the least of its parts, and what is more, it always remains the same machine" (GP IV:482). Note that this characterization constitutes the main difference between two *types* of machines. Furthermore, this characterization applies both to the internal structure of a natural machine, so that all of its parts are machines, and to its development, so that it remains the same machine through its various states. Both the unity of a natural machine at a particular time and its identity over time are preserved (throughout the changes it undergoes).

Leibniz's view concerning the unity and identity of a natural machine, in contrast to the lack of unity and identity of an artificial one, is confirmed in the sequel to the passage cited earlier:

> In addition, by means of the soul or form there is a true unity corresponding to what is called *the self* in us. Such a unity could not occur in the machines made by a craftsman or in a simple mass of matter, however organized it may be; such a mass can only be considered as an army or a herd, or a pond full of fish, or like a watch composed of springs and wheels. (AG 142)[9]

Leibniz draws a sharp distinction here: artificial machines are to be understood on the model of things that lack true unity—namely, aggregates, however organized they may be or however many living beings they may contain. By contrast, natural machines are to be understood on the model of things that have true or substantial unity. Even if it involves infinitely many states and infinitely many machines, a natural machine is regarded as a true unity. By 1695, Leibniz was well equipped with this fundamental distinction between substances and aggregates, which he develops and defends in the second part of his correspondence with Arnauld (1686-87).[10] While a

---

[9] "De plus, par le moyen de l'âme ou forme, il y a une véritable unité qui répond à ce qu'on appelle moi en nous; ce qui ne sauroit avoir lieu ni dans les machines de l'art, ni dans la simple masse de la matière, quelque organisée qu'elle puisse être, qu'on ne peut considérer que comme une armée ou un troupeau, ou comme un étang plein de poissons, ou comme une montre composée de ressorts et de roues" (GP IV:482).

[10] "La substance demande une véritable unité. . . . Tout être par agrégation suppose des êtres doués d'une véritable unité, parce qu'il ne tient sa réalité que de celle de ceux dont il est

substance has a true unity, an aggregate, which is a collection of substances, does not. The unity of an aggregate requires a mental act—that is, the very aggregation of its constituents into a single group (such as the individual sheep into a herd, stones into a pile, soldiers into an army, birds into a flock, etc.). Such a union is not natural but a result of a mental operation of unification, namely that of perceiving a plurality of things (sheep, fish, stones, soldiers) together as a single group (A 6.4:627).[11] I have to stress that these are just analogies and examples to illustrate something that cannot in fact be visualized—namely, the difference between a true and natural unity and an artificial one. Samuel Levey has nicely articulated this distinction, as follows:

> Substances—the fundamental beings in the world's inventory—have a "unity" or "one-ness" that is intrinsic to them. They are beings of which it is unqualifiedly true to say: here is some *one* thing. Leibniz describes beings that have this sort of unity as "true unities" or as "substantial unities" or "one *per se*."
>
> By contrast, aggregates do not have this sort of unity. Aggregates may be aggregates *of* substances, but they are not themselves true unities, one *per se*. Rather, they are only multitudes, pluralities, heaps. Whatever unity an aggregate may enjoy is extrinsic to it, an aspect of the way its constituents are represented in thought to some mind. In truth, an aggregate . . . is not one thing at all but only so many distinct things: those things that are being aggregated together.[12]

This fundamental difference in Leibniz between a unity (a substance) and a plurality (of substances) in relation to infinity will be considered later. What I would like to stress here is that artificial machines fall on the aggregate side of the divide and that natural machines fall on the substance side of the divide. Thus, according to Leibniz, a natural machine is one thing and an artificial machine is a composition (or, at any rate, a multitude) of many things. Yet, it is not at all clear how Leibniz can account

---

composé, de sorte qu'il n'en aura point du tout, si chaque être dont il est composé est encor un être par agrégation. . . . S'il y a des agrégés de substances, il faut bien qu'il y ait aussi des véritables substances dont tous les agrégés soient faits. . . . Il n'y a point de multitude sans des véritables unités. Pour trancher court, je tiens pour un axiome cette proposition identique qui n'est diversifiée que par l'accent, que ce qui n'est pas véritablement un être, n'est pas non plus véritablement un être" (LR 165; see also GP II:164–65).

[11] For Leibniz's notion of aggregate and its peculiar (semi-mental) sense of unity, see P. Lodge, "Leibniz's Notion of an Aggregate," *British Journal for the History of Philosophy* 9, no. 3 (2001): 467–86; Nachtomy (2007a, ch. 8).

[12] S. Levey, "On Unity and Simple Substance in Leibniz," *Leibniz Review* 17 (2007): 64.

for (and justify) this distinction, if the sole difference between them is that a natural machine remains a machine to the least of its parts.

Before addressing this question more directly, let me briefly return to Leibniz's motivation for drawing the distinction and to some of the roles it plays in his metaphysics. On this point, let me refer to Fichant's article, "Leibniz et les machines de la nature." Fichant has emphasized that,

> The concept [of a natural machine] is introduced as a way to limit the pretensions of an integral mechanism, which "in confounding natural and artificial things," has reduced natural phenomena to effects of machines closely analogous to machines of human artifice.[13]

In contrast to Descartes, according to Leibniz, the difference between "the least productions and mechanisms of divine wisdom and the greatest works of human art" is not one of degree but rather one of kind. Leibniz attempts to limit the claims of the all-encompassing Cartesian mechanistic program by seeking not only to draw a line between divine and human production but also between living and nonliving things. As it turns out, for Leibniz, this also implies drawing a line between active (or animate) and nonactive (or inanimate) things and, likewise, between truly existing things (which he typically identifies with substances) and well-founded phenomena (which he typically identifies with aggregates). At the same time, Leibniz's characterization of both divine creation and human production in terms of machines is meant to meet his conviction that things can be described both mechanically, by appealing to efficient causes, and teleologically, by appealing to final causes.[14] In other words, Leibniz thinks that the physiological aspects of the living world should be described mechanically. But this does not permit ignoring its metaphysical grounds—grounds

---

[13] Fichant (2003, 1). See also *Monadologie* 74: "a kind of divine machine which infinitely surpasses all artificial automats."

[14] For Leibniz, and for some of his contemporaries, a machine is understood not only in terms of efficient causes but also in terms of final causes, as an instrument—a point to which I will return in the final section. "In each machine, one has to take into consideration at once its functions or its end and the mode of operation or the means by which the author of the machine sought its end"; "In omni Machina spectandae sunt tum functiones ejus, sive finis, tum modus operandi, sive quibus mediis autor machinae suum finem sit consecutus," in E. Pasini, *Corpo et funzione cognitivi in Leibniz* (Milan: Franco Angeli, 1996), 212. Also, "Machina autem omnis a finali causa optime definitur, ut in explicatione partium deinde appareat, quomodo ad usum destinatum singulae coordinentur" (217–18); "Any machine, moreover, is best defined in terms of its final cause, so that in the description of the parts it is therefore apparent in what way each of them is coordinated with the others for the intended use" Smith (2011, app. 3, 290). See also Arthur (2014, 73).

that cannot be described in mechanical terms. Rather, this ground relates to the features of activity, unity, and infinity that we have traced in the previous chapter.[15]

We can now better appreciate the significance of the distinction for Leibniz. Clearly, a lot hangs on it. Not only does it serve to distinguish between divine creation and human production but it also distinguishes between animate and inanimate things, as well as reconstructs a new model of the living world in which the Aristotelian notions of entelechy, form, and *telos* would play a central role—yet a role consistent with a mechanistic description of physiology and other life-related phenomena.[16]

6.4 ON THE COHERENCE OF LEIBNIZ'S DISTINCTION

At this point, one must wonder whether Leibniz's distinction makes sense. In other words, let us examine whether Leibniz has the resources to justify the distinction and the important roles it is to play in his metaphysics. In particular, we need to examine whether the distinction can indeed differentiate between divine creations and human artifacts, true unities and aggregates, and animate and inanimate things—and we need to examine whether the resulting concept of a natural machine is a coherent one. The most striking difficulty concerning the notion of a natural machine is this. As Fichant observes, the central characteristics of a natural machine are (1) that its composition extends to infinity; and (2) that it is a true unity.[17] And it is precisely the conjunction of these two traits (composition to infinity and true unity) that is difficult to grasp. In other words, it is difficult to grasp how Leibniz considers infinitely many machines within machines as one substantial thing. (Note that we have already considered a more general variant of the question in chapter 4.)

It is clear that, according to Leibniz, a natural machine is supposed to have a substantial unity that an artificial machine would lack: a natural machine is like a sheep; an artificial machine is like a herd. We know that, for Leibniz, the individual sheep has natural and substantial unity, which the herd, the army, or the clock does not. But the picture is more complicated, in two respects. On the one hand, an artificial machine, too, has substance-like, sheep like constituents. It is, in brief, an aggregate

---

[15] I develop this point in terms of the distinction between the source of life and the phenomena of life in section 8.5.
[16] In addition, as Fichant has argued, "in the *Système Nouveau*, Leibniz is concerned ... with a structural and ontological characterization of natural machines in an attempt to give bodies the reality of a substance" (Fichant 2003, 7). According to Fichant, this strategy is the basis for a realistic interpretation of substance that extends well into the Monadological period, which is regarded by many as idealistic.
[17] "Cette différence se marque à deux traits: l'infinité de composition, garante d'indestructibilité, et l'unité véritable, fondement de substantialité" (Fichant 2003, 2).

of *substances*. On the other hand, each sheep or a natural unity itself consists of other sheeplike, substance-like, creatures.[18] The challenge, then, is to distinguish between an artificial machine and a natural machine, both seemingly consisting of infinitely many natural machines. Clearly, Leibniz's distinction must be rather nuanced, if it can account for this difference.

In a number of texts, Leibniz offers the following distinguishing mark between these two kinds of machines: while a natural machine is infinite, an artificial machine is finite. In the "New System," Leibniz states that "the machines of nature have a truly infinite number of organs, and are so well supplied and so resistant to all accidents that it is impossible to destroy them" (GP IV:482; AG 142). Leibniz adds that a natural machine "is made up of an infinity of entangled organs."[19]

These passages suggest that, according to Leibniz, the distinctive feature of a natural machine is that it has infinitely many organs. Yet this cannot be all there is to his distinction. In fact, praising the subtlety of natural machines is not far from what Descartes says (with the important qualification that Leibniz is committed to *infinite* subtlety, to which Descartes is not). Taken at face value, however, Leibniz's claim that a natural machine "is made up of an infinity of entangled organs" cannot account for the difference between artificial and natural machines. The reason is that, as we have observed, an artificial machine would involve infinitely many organs as well. If an artificial machine consists of infinitely many natural machines, it would also have infinitely many organs and in this sense would be indistinguishable from a natural one.

At the same time, it is clear in these passages that, rather than being an obstacle, Leibniz sees the composition to infinity as what *guarantees* the unity and indestructibility of a natural machine. But the mere infinity of organs or machines cannot account for this alleged unity and indestructibility. There are two reasons for this: First, a mere infinity of organs does not provide unity but, if anything, has multiplicity

---

[18] "Dans les corps je distingue la substance corporelle de la matiere, et je distingue la matiere premiere de la seconde. La matiere seconde est un aggregé ou composé de plusieurs substances corporelles, comme un troupeau est composé de plusieurs animaux. Mais chaque animal et chaque plante aussi est une substance corporelle, ayant en soy le principe de l'unité, qui fait que c'est véritablement une substance et non pas un aggregé. Et ce principe d'unité est ce qu'on appelle Ame ou bien quelque chose, qui a de l'analogie avec l'ame. Mais outre le principe de l'unité la substance corporelle a sa masse ou sa matiere seconde, qui est encor un aggregé d'autres substances corporelles plus petites, et cela va à l'infini" (draft letter to Thomas Burnett, 1699; AG 289; GP III:260).

[19] "Moreover, a natural machine has the great advantage over an artificial machine, that, displaying the mark of an infinite creator, it is made up of an infinity of entangled organs. And thus, a natural machine can never be absolutely destroyed just it can never absolutely begin, but it only decreases or increases, enfolds or unfolds, always preserving in itself some degree of life [*vitalitas*] or, if you prefer, some degree of primitive activity [*actuositas*]" ("On Body and Force, Against the Cartesians," May 1702; AG 253; GP IV:396).

and infinite divisibility. As we have seen, for Leibniz, "an infinity of things is not one whole" (A 6.3:503). Second, as far as we know, Leibniz cannot accept without qualification an infinity of organs as making up one whole because he rejects the notion of an infinite number as a contradictory notion.[20] Thus, if Leibniz's distinction is supposed to turn on the infinite versus a finite *number* of organs, it does not seem to provide a workable solution. Instead, it would seem to render his notion of a natural machine one that not only lacks unity but also lacks sense—that is, an altogether contradictory idea.

It might prove more promising to attend carefully to Leibniz's repeated claim regarding the infinite number of organs in a natural machine. In particular, I will try to clarify what Leibniz means by "organ" in this context. My conjecture—developed later—is that this might be a different way of expressing the view that a natural machine remains a machine to the least of its parts in the sense that each organ serves a certain function. Likewise, let us note that Leibniz suggests that what extends to infinity is not the *number* of organs or parts but, rather, the whole machine, as including more machines within machines to infinity. It goes without saying that we also need to pay careful attention to what Leibniz means by "infinite" in this context. In order to clarify the sense in which he employs the notion of infinity, I will now lay out two ways to interpret Leibniz's claim that a machine remains a machine in the least of its parts: a structural reading and a functional reading. I begin with the former.

6.5 A MACHINE TO THE LEAST OF ITS PARTS: A STRUCTURAL READING

The structural reading of Leibniz's notion of a natural machine is suggested by passages such as the following:

> the machines of nature being machines to the least of their parts are indestructible, due to the envelopment of a small machine in a larger one, to infinity. (GP VI:543)[21]

In the following passage, from a 1704 letter to Lady Masham, Leibniz claims that, in a natural machine, the composition goes to infinity or, more precisely, that the subtlety of its artifice extends to infinity:

---

[20] For details on this issue, see chapters 1 and 2, this volume.
[21] See also: "[Le] corps est organique quand il forme une manière d'automate ou de machine de la nature, qui est machine non seulement dans le tout, mais encore dans les plus petites parties qui se peuvent faire remarquer" (PNG §3). See also *Monadologie* 67–70.

> I define an organism or a natural machine, as a machine each of whose parts is a machine, and consequently the subtlety of its *artifice* extends to infinity. (GP III:356, my emphasis)

According to the reading I propose here, what extends to infinity is not the *number* of organs or machines but, rather, the whole *structure* of a natural machine, which involves machines within machines to infinity. In previous publications, I called this feature the nested structure of Leibnizian individuals.[22] My suggestion here is that the structure of a natural machine, with its intricate relations between the machines nested in it, develops *ad infinitum*, while that of an artificial machine does not. It is in this sense, I suggest, that an artificial machine does not remain a machine to the least of its parts. While the number of machines within this structure is clearly not finite, we cannot also say that it involves an infinite number of machines (which would be a contradiction, according to Leibniz), but that the machine's structure extends to infinity. Before I turn to exemplify this point, let us consider an objection.

One might object that this only implies that we need to count structures instead of organs. But, if so, the contradiction would arise with an infinite number of structures as well. Let me clarify. While the structure of a natural machine might include many sub-structures, the point is that there is one structure corresponding to the whole machine—and that that structure might involve many nested machines as constitutive elements. For that reason, too, a natural machine is not nested in a larger machine. It has, so to speak, an upper bound, which as we shall see is expressed by the domination relation. Intuitively, the sheep is not nested in the herd. The herd is not a natural machine but, rather, an aggregate of natural machines; each of them is a complete entity.

Leibniz's picture of nestedness to infinity is not a simple containment or inclusion of one thing inside another. This can be seen in a passage in the "New Essays" in which Leibniz evokes the image of the Harlequin. While this image might be misleading, notice that Leibniz is *denying* that the Harlequin provides a good model for the richness of natural subtlety:

> it is as if someone tried to strip Harlequin on the stage but could never finish the task because he had on so many costumes, one on top of the other; though the infinity of replications of its organic body which an animal contains are not as alike as suits of clothes, and nor are they arranged one on top

---

[22] See Nachtomy (2007a, ch. 9).

of another, since nature's artifice is of an entirely different order of subtlety. (NE 2.7.42, p. 329)[23]

Leibniz does not clarify here what he has in mind when he says that "nature's artifice is of an entirely different subtlety" from that of human production. I suggest that the difference between human-made machines and natural/divine ones is this: while an artificial machine might also have infinitely many parts, a natural machine has an internal structure that extends to infinity. More important still (and I will try to illustrate this later), a natural machine, while infinite in structure, is essentially one, and therefore, its infinity must be compatible with its having true unity. Given Leibniz's distinctions explored in earlier chapters, we can call the infinity that pertains to a natural machine an infinity presupposed by unity, and the infinity that pertains to an artificial machine a syncategorematic infinity. As we have seen in the previous chapter, the kind of infinity that applies to unities characterizes beings, while the kind of infinity that applies to a multiplicity characterizes aggregates. This distinction agrees with our observation that a natural machine belongs to the category of substance while an artificial machine belongs to the category of aggregates.

Let me now try to illustrate this difference. I will do so by using the concept of a law of a series—a concept whose importance for Leibniz we have already discussed several times. But here it takes a curious application. Let us think of a natural machine as having a fractal-like structure—that is, a structure defined by a unique rule of generation, whose continuous application produces an infinite structure, such that each of its parts has the same structure as the whole. While the analogy with a fractal structure might seem anachronistic, let us attend to what Leibniz writes to Des Bosses in 1706:

> When I say that there is no part of matter that does not contain monads, I illustrate this with the example of the human body or that of some other animal, any of whose solid and fluid parts contain in themselves in turn other animals and plants. And this, I think, must be said again of any part of these living things, and so on to infinity. (March 11, 1706; GP II:305–306; LDB 35)

---

[23] "[C]'est ... comme Arlequin qu'on voulait dépouiller sur le théâtre, mais on n'en put venir à bout, parce qu'il avait je ne sais combien d'habits les uns sur les autres: quoique ces réplications des corps organiques à l'infini, qui sont dans un animal, ne soient pas si semblables ni si appliqués les unes aux autres, comme des habits, l'artifice de la nature étant d'une tout autre subtilité" (NE 2.7.42; GP V:309).

To an objection that this view seems to imply an infinitesimally small being, Leibniz responds,

> I shall use an analogy.²⁴ Imagine a circle; in it draw three other circles which are the same size and as large as possible, and in any new circle and in the space between circles again draw the three largest circles of the same size which are possible. Imagine proceeding to infinity in this way: it does not follow that there is an infinitely small circle, or that there is a center having its own circle, in which (contrary to the hypothesis) no other is inscribed. (March 11, 1706; GP II:305–306; LDB 35)

It is easy enough to illustrate the geometrical analogy Leibniz draws here. Similar illustrations appear elsewhere in his writings (e.g., A 6.3:525; DSR 86–87). As it turns out, Leibniz's example corresponds to the contemporary definition of a fractal. It is produced by a simple generation rule and each of its parts is homomorphic to the whole.²⁵ Notice, too, that such a fractal structure presents a situation Leibniz is fond of describing poetically as follows: "It is all as here, everywhere and always" (*C'est tout comme ici, partout et toujours*), which he claims to be one of the central principles of his philosophy (see especially the letter to Sophie Charlotte of May 8, 1704; GP III:343–48).²⁶

Leibniz may be overstating his case a little when he claims that "all is as here, everywhere and always." While each of the internal structures in a fractal is the same as the whole with respect to its structure, there is also an important difference. If we take Leibniz's principle of considering the whole method of production of a given thing, we see that there are differences between these structures. For example, the place of each structure within the whole structure is obviously different: while each structure is the same, the place each component has within the whole differs.

---

²⁴ I should note that some commentators (e.g., Chazerans 1991) who are developing this analogy are not attending to the fact that the geometrical analogy, which they call the schema of *emboîtement*, does not come right after the passage cited. In between, there is a complex discussion not only about matter but also about machines, entelechies and their complex relations. In fact, it is not obvious which passage Leibniz attempts to exemplify with the analogy. What he says immediately before the passage just cited is this: "Yet you see that it should not be concluded from this that an infinitely small portion of matter (such as does not exist) must be assigned to any entelechy, although we routinely jump to such conclusions."

²⁵ It is also worth recalling a simpler example Leibniz refers to on various occasions—namely, a line—and considering each line which is a part of it, such that each part itself contains infinitely many lines, just like the whole.

²⁶ For more on the way in which Leibniz uses this principle, see P. Phemister, "'All the time and everywhere everything's the same as here': The Principle of Uniformity in the Correspondence between Leibniz and Lady Masham," in *Leibniz and His Correspondents*, ed. P. Lodge (New York: Cambridge University Press, 2004), 193–213.

Let us now apply Leibniz's analogy for our current purposes. In this analogy, an artificial machine would be seen as a collection of fractals. A natural machine would be a single fractal that includes infinitely many subfractals as its intrinsic and proper constituents.[27] Note that, in this illustration, a natural machine would remain a machine to the least of its parts, while at the same time the whole machine would remain one single machine. An artificial machine, however, does not preserve this structure to infinity, nor is it, for this very reason, a truly single unit—not at any given moment and not over time, even if it is composed of or entails such machines. On this model, it seems, we can maintain Leibniz's point that the distinction between artificial and natural machines coincides with the distinction between a true unit (that is, a substance) and a collection of them (that is, an aggregate).

In addition, we have seen that Leibniz defines an individual substance in terms of its individual law of generation—that is, the law of the series. Drawing on the fractal analogy as exemplifying how such a law of generation can produce a structure that develops to infinity, we can suggest that a natural machine can be defined as including an infinity of machines and as having a nested structure to infinity, in the sense that its law of production can be seen as including sublaws as essential constituents (but not parts) of it.[28]

An artificial machine, however, is not constituted in this way. Rather, it is seen as a collection *of* such individuals, not as an individual that makes up one whole. If this is correct, the distinction between artificial and natural machines turns, strictly speaking, on the question of unity, or more precisely, it turns on the appropriate conjunction of unity and infinity. In fact, Leibniz's claim for the composition to infinity of a natural machine suggests that its individuation is due to a single law or a single program of action. On this reading, a natural machine turns out to be one *thing*, while an artificial machine, being an aggregate, turns out to be a compositional product, or a collection of many things. Thus we can see why Leibniz likens natural machines to substances and artificial machines to aggregates.

Let us now examine how this reading fits with the distinction between divine creation and human production. According to Leibniz, God creates complete

---

[27] The idea of using a fractal analogy to exemplify the distinction between a composed substance and an aggregate has been proposed (though in a very loose and imprecise way) in an article by J-F. Chazerans, "La substance composée chez Leibniz," *Revue Philosophique de France et de l'étranger* 1 (1991): 47–66.

[28] At the same time, it is very difficult to see how this mathematical ideal of a fractal structure may be reconciled with the structure of living beings as we know them. Thus, there is no straightforward analogy between the presumed nested elements of an animal and the organs of its body, such as heart, limbs, etc. In sections 8.4 and 8.5, this volume, I explain some of the reasons for this gap between Leibniz's metaphysical commitments and the phenomena of life. I stress that Leibniz's thesis here should be taken in a metaphysical sense that remains quite detached from the empirical phenomena.

individual substances—the whole world (space, time, extended bodies, and whatever else there is) supervenes on the existence of individual substances. Furthermore, such substances are individuated by their complete concepts, which are conceived in God's mind as possibles before their creation. I suggested that such a concept should be defined not as a set of predicates but through the law that generates a unique structure of predicates in God's mind.[29] The main reason for defining the concept of an individual in this way is that such a genetic definition (via a generative rule) can capture the infinite character of a Leibnizian individual in a consistent way. Otherwise, if we define it simply as a set or as a collection of infinite predicates, it would fall prey to the contradictions of infinite number. The definition of a complete concept in terms of its law of production aims to capture Leibniz's characterization of an individual substance as having an infinite structure, as well as informing its development upon creation.

These observations further clarify Leibniz's identification of natural machines with divinely created individual substances. Surely such an infinite structure cannot be assembled by humans; it can only be created as a *complete* entity by God. A natural machine cannot be assembled from parts. Rather, it can only be brought about by an act of creation—that is, a supernatural event constituting the natural world by realizing a variety of natural machines. As a natural machine cannot be naturally produced, it cannot be naturally destroyed, either. For it would exist as long as it would act, and the very being of natural machines is related to their activity and force (see "On Nature Itself"). Thus, we see that the indestructibility of a natural machine goes hand in hand with the fact that it cannot be produced, but can only be (supernaturally) created (and annihilated) by God.[30]

Leibniz makes it very clear that the indestructibility of natural machines is related to their composition to infinity. As he explains to Des Bosses:

> Whoever reflects on the doctrine of the conservation of animals, must also consider, as I have shown, that there are infinite organs in the body of an animal, some enfolded in others; *and from this it follows* that an animated machine, and in general a machine of nature, is absolutely not destructible. (Leibniz to Des Bosses, March 11, 1706; LDB 37, my emphasis)

---

[29] Nachtomy (2007a).

[30] "Quand aux Mouvemens des corps celestes, et plus encore quant à la formation des plantes et des animaux, il n'y a rien qui tienne du miracle, excepté le commencement des ces choses. L'organisme des animaux est un mechanisme qui suppose une préformation divine: ce qui en suit, est purement naturel, et tout à fait mechanique" (GP VII:417–18).

The beginning of this passage[31] shows that Leibniz connects the lawfulness of natural machines (created by God), their nested structure, and their natural indestructibility. He restates this point in his "Consideration on the Principles of Life":

> What again reveals to us the marvels of the divine artifice, ... is that natural machines, being machines to their least parts, are indestructible, thanks to the enveloping of a small machine within a larger one to infinity.[32]

As Leibniz says here, natural machines are indestructible owing to the envelopment of a small machine by a greater one, *ad infinitum*. My suggestion is that natural machines are indivisible units in the sense that they are defined and informed by a single rule of generation, compatible with their having an infinitely complex structure, such as an infinite series or a fractal-like structure.

As we have already seen, the infinite structure of a natural machine is consistent with its being a divine, and law-governed, creation. These strands come together in the following passage from the PNG, section 3:

> And this body [of the central monad] is organic when it forms a kind of automaton or a natural machine, which is not only a machine as a whole, but also in the smallest parts that can be distinguished. And since everything

---

[31] "As to my claim that the soul and the animal do not perish, I shall again explain it with an analogy. Imagine an animal as a drop of oil and the soul as some point in the drop. If the drop is now divided into parts, the point will exist in one of the new drops, since any part in turn is transformed into a spherical drop. In the same way, the animal will survive in that part in which the soul remains and which best agrees with the soul itself. And just as the nature of the liquid in any fluid aims at sphericity, so the nature of the matter constructed by the wisest author always aims at order or organization. From this it follows that neither souls nor animals can be destroyed, although they can be diminished and concealed, so that their life does not appear to us. And there is no doubt that in generation, as also in corruption, nature obeys certain laws, for nothing of divine workmanship is lacking in order. Moreover, whoever reflects on the doctrine of the conservation of animals, must also consider, as I have shown, that there are infinite organs in the body of an animal, some enfolded in others; and from this it follows that an animated machine, and in general a machine of nature, is [not] absolutely destructible" (LDB 35–37).

[32] "Ce qui nous découvre encore des merveilles de l'artifice divin, où l'on n'avait jamais pensé: c'est que les machines de la nature étant machines jusque dans leurs moindres parties, sont indestructibles, à cause de l'enveloppement d'une petite machine dans une plus grande à l'infini" (GP VI:543). See also this passage: "la matière arrangée par une sagesse divine doit être essentiellement organisée partout, et qu'ainsi il y a machine dans les parties de la machine naturelle à l'infini, et tant d'enveloppe et des corps organiques enveloppés les uns dans les autres, qu'on ne saurait jamais produire un corps organique tout à fait nouveau" (GP VI:539–46).

is connected because of the plenitude of the world, . . . it follows that each monad is a living mirror or a mirror endowed with internal action, which represents the universe from its point of view and is as ordered as the universe itself. And the perceptions in the monad arise from one another by the laws of appetites, or by the laws of the *final causes of good and evil.* (PNG §3, AG 207)[33]

This passage is remarkable in clarifying the conditions under which a body is considered organic and for tying together the nested structure of a natural machine with its internal law of action (a law that regulates its perceptions). In the next chapter, we shall examine the notion of a living mirror in some detail. Presently, it is more important to point out that this passage reveals a close connection between the internal law of action and the final causality involved in the internal perceptions of a natural machine, which I now turn to consider.

6.6 A MACHINE TO THE LEAST OF ITS PARTS: A FUNCTIONAL READING

Let me now examine another way in which the subtle distinction between natural and artificial machines may be understood— namely, by emphasizing a functional reading of the notion of machine (and of machines within machines). This sense of machine is related to its traditional source in the notion of *organon*, seen as an instrument for performing a certain function or use. In texts from the 1680s, Leibniz is very explicit about the functional role of machines. He writes, "[i]n each machine, one has to take into consideration at once its functions or its end and the mode of operation or the means by which the author of the machine sought its end."[34] He is even more explicit in the following passage: "The best way to define a machine is by its final cause, in a way that each of its parts would appear [in the explication of its parts] to be coordinated with the other by its designated [*destinatum*] usage."[35]

The functional reading of Leibniz's notion of a natural machine is also clearly expressed in the *Monadologie*, where he writes,

---

[33] "Et ce corps [de la Monade Centrale] est organique, quand il forme une manière d'Automate ou de Machine de la Nature, qui est Machine non seulement dans le tout, mais encore dans le plus petites parties qui se peuvent faire remarquer. . . . Et les perceptions dans la Monade naissent les unes des autres par les lois des Appétits, ou des causes finales du bien et du mal."

[34] "In omni Machina spectandae sunt tum functiones ejus, sive finis, tum modus operandi, sive quibus mediis autor machinae suum finem sit consecutus" (Pasini 1996, 212); see app. 3 in Smith (2011).

[35] "Machina autem omnis a finali causa optime definitur, ut in explicatione partium deinde appareat, quomodo ad usum destinatum singulae coordinentur" (Pasini 1996, 217–18).

[t]hus each organized body or a living being is a kind of divine machine or a natural automaton, which infinitely surpasses artificial automata. For a machine constructed by man's art is not a machine in each of its parts. For example, the tooth of a brass wheel has parts or fragments which, for us, are no longer artificial things, and no longer have any marks to indicate the machine for whose use the wheel was intended. But natural machines, that is, living bodies, are still machines in their least parts, to infinity. That is the difference between nature and art, that is, between divine art and our art. (*Monadologie* 64; AG 221)[36]

Here, it seems that "to remain a machine to the least of its parts" means that in a natural machine each of its parts contributes to the end of the whole machine by performing a certain function. An artificial machine is invested with human purposes and with the use humans make of it. Yet, at a certain level of its internal structure, these purposes come to an end. The machine as a whole has a purpose but not each of its constituents, or, more precisely, not each of its constituents is made of such machines to infinity. The cog, for example, has a function within the machine, and in this sense it, too, is a machine; the dents on the wheel have a function as well, but this functional structure does not continue to the fragments of the dents. At this point, the functional chain terminates. And this would explain why it is seen as an artificial machine rather than as a natural one.[37]

---

[36] "[U]ne Machine, faite par l'art de l'homme, n'est pas Machine dans chacune de ses parties, par exemple le dent d'une roue de loton a des parties ou fragments, qui ne nous sont plus quelque chose d'artificiel et n'ont plus rien qui marque de la Machine par rapport à l'usage où la roue était destinée. Mais les Machines de la Nature, c'est à dire les corps vivants, sont encor des machines dans leurs moindres parties jusqu'à l'infini. C'est ce qui fait la différence entre la Nature et l'Art, c'est à dire entre l'art Divin et le Notre" (GP VI:618).

[37] It is worth reflecting on the similarity between Leibniz's formulation here and Kant's formulation of the distinction between nature and art. In his *Critique of the Power of Judgment*, Kant writes: "In such a product of nature each part is conceived as if it exists only **through** all the others, thus as if existing **for the sake of the others** and **on account** of the whole, i.e., as an instrument (organ), which is, however, not sufficient (for it could also be an instrument of art, and thus represented as possible at all only as an end); rather, it must be thought of as an organ that **produces** the other parts (consequently, each produces the others reciprocally), which cannot be the case in any instrument of art, but only of nature, which provides all the matter for instruments (even those of art): only then and on that account can such a product, as an **organized** and **self-organized** being, be called a **natural end**" (I. Kant, *Critique of the Power of Judgment*, in *Cambridge Edition of the Works of Immanuel Kant*, ed. Paul Guyer and Allen Wood [Cambridge: Cambridge University Press, 2000], 245; and I. Kant, *Gesammelte Schriften*, German edition, ed. Königlich Preussische Akademie der Wissenschaften [Berlin: Walter de Gruyter, 1902–], 5:373–74).

By contrast, a natural machine expresses God's purposes and designs and, in this respect, too, it is of a different category: in a natural machine, the functional and machine-like structure go to infinity while in an artificial machine they come to an end. In this sense, the one is infinite while the other is not.

Evidently, according to Leibniz, there is nothing created by God that does not fulfill a certain function. More precisely, everything is created with respect to its function or end in the world. This is a familiar Leibnizian theme. The reason God creates a certain individual is its contribution to the harmony and perfection of the world. Note that, in this functional sense of nestedness to infinity, the functional chain, or the chain of final causes need not at all be seen in a physical or even structural sense of *emboîtement*, or of machines within machines in the literal sense explored prior. What is crucial here is only that, at every level, each part or constituent serves a function with respect to the other constituents and with respect to the main (dominating) *telos* of the whole. Such a model of functional relations may well be illustrated as a circular rather than linear (or serial) progress to infinity.[38]

As we have seen, this system of functional relations does not apply to human production in the same way. Even if the cog is made up of other things, and ultimately these things are going to be living things, they are not functionally related as organs are typically related to the whole organism. In a living animal, the constituents are seen as inseparable and as inseparably individuated from the whole structure and *telos* of the animal. In this respect, my liver is not like a cog on my bicycle, whether or not the technology for their replacement exists. On this reading of Leibniz, what distinguishes between the natural and the artificial is precisely that the network of functional and teleological relations is never ending—any natural thing, however small or insignificant, serves a certain function in a well-ordered system of ends. But this is not the case in an artificial machine, whose series of functions does come to an end.

In this vein, Leibniz draws a distinction between the ends of machines, which are proper and interior to them, and the ends of aggregates, which are the result of the relations between different machines. In the controversy with Stahl, this distinction is made very explicitly between particular final causes that Leibniz ascribes to natural machines and general final causes that he ascribes to the concurrence between them:

> we concede that there is a great difference between machines and aggregates or masses, since machines have ends and effects through the force of their

---

[38] Once again, one is reminded here of Kant's formulation that, in a natural product, each organ is both means and end: "An organized product of nature is that in which everything is an end and reciprocally a means as well. Nothing in it is in vain, purposeless, or to be ascribed to a blind mechanism of nature" (Kant 2000, 247–48; Kant 1902–, 5:376).

structure, whereas the ends and effects of aggregates arise from a series of concurrent things and thus from the convergence of diverse machines. (LSC 249)[39]

6.7 CONCLUSION

I have presented two ways to read Leibniz's characterization of a natural machine as remaining a machine to the least of its parts—one structural, suggested by Leibniz's fractal analogy, and one functional, suggested by examples such as in *Monadologie* 64. Let me briefly touch on the question of their relations. The question arises whether the functional and structural readings are compatible, or whether they exclude one another. In other words, are these readings complementary, so that the one contributes to the other, or are they independent from one another?

My suggestion is, briefly, this: the internal law of the structure of a natural machine expresses the unique *telos* of the whole machine, as well as the machines nested in it. Thus, through the generative law, the structural and functional aspects of a natural machine are not only compatible but also complementary. For this reason, both structural and functional considerations are essential to Leibniz's notion of natural machine.

Finally, let me address the following question: Does the same kind of infinity apply to both natural and artificial machines? And, if not, what kind of infinity applies to natural machines, and what kind to artificial machines? While the distinction seems very subtle, it involves, I suggest, an employment of a different kind (or degree) of infinity. The infinity involved in Leibniz's notion of an artificial machine is a syncategorematic one that applies to the multiplicity of parts or constituents that make up or compose an artificial machine as an accidental unity. On the other hand, the infinity involved in a natural machine pertains to a nondivisible unity—one that is not (and cannot be) composed. As I will expand on in the next chapter, it is an infinity that characterizes being and unity. And yet it is not the absolute infinity that applies to God alone but, rather, an infinity of a particular being, which is also limited. The distinctions between these different degrees of infinity are further explored in the next two chapters.

---

[39] "Interim concedimus magnum esse discrimen inter machinas et aggregata massasque, quod machinae fines et effectus habent vi suae structurae, at aggregatorum fines et effectus oriuntur ex serie rerum concurrentium atque adeo ex diversarum machinarum occursu, qui etsi etiam sequatur divinam destinationem, plus tamen minusque manifestae coordinationis habet" (LSC 248).

7

LIVING MIRRORS AND MITES
LEIBNIZ AND PASCAL

7.1 INTRODUCTION

In this chapter, we again consider the question of infinity and its application to living beings, but in a specific context: the chapter focuses on a dense, rich text in which Leibniz comments on fragment 22 of Pascal's *Pensées* in the Port Royal Edition (currently indexed by Lafuma §199). The importance of this text has been already noted by Gerhardt in 1891[1] and then by Baruzi in 1907;[2] it was reedited by Grua in 1948 under the charming title "Double infinité chez Pascal et Monade," which facilitated further commentaries by Guitton,[3] Serres,[4] Naërt,[5] McKenna,[6] and Carraud,[7] among others. The text was dated by Grua to sometime after 1695.[8] It has received a fair amount of commentary in French, but so far none in English. In this text, Leibniz responds to Pascal's employment of the infinitely large and infinitely small and, more specifically, he responds to Pascal's use of infinity to describe living beings through the example of a mite (*ciron*). The relation of this text to Leibniz's use of infinity as a defining feature of living things has been largely overlooked. In contrast to Pascal's mite, Leibniz employs for the first time in his career the image of a living mirror (*miroir vivant*).

---

[1] C. I. Gerhardt, "Leibniz und Pascal," in *Sitzungsberichte der Königlichen Akademie der Wissenschaften zu Berlin*, vol. 2 (Berlin: Verlag der Königlichen Akademie der Wissenschaften, 1891).
[2] J. Baruzi, *Leibniz et l'organisation religieuse de la terre* (Paris: Félix Alcan, 1907), 224–31.
[3] J. Guitton, *Pascal et Leibniz* (Paris: Aubier, 1951).
[4] M. Serres, *Le système de Leibniz et ses modèles mathématiques*, vol. II (Paris: Presses Universitaires de France, 1968), 719–21.
[5] E. Naërt, "Double infinite chez Pascal et Monade," *Studia Leibnitiana* 17, no. 1 (1985): 44–51.
[6] A. McKenna, *De Pascal à Voltaire. Le rôle des Pensées de Pascal dans l'histoire des idées entre 1670 et 1734*, 2 vols. (Oxford: Voltaire Foundation, 1990).
[7] Carraud studies the philosophical relations between Leibniz and Pascal in some detail. See V. Carraud, "Leibniz lecteur des Pensées de Pascal," *XVIIe siècle* 2 (1986): 107–24; and V. Carraud, *Pascal et la philosophie* (Paris: Presses Universitaires de France, 1992).
[8] Grua 553–55.

Through this image, Leibniz seeks to show that active perception of its own structure (that develops to infinity) allows each living being, no matter how small, to represent and mirror the whole universe. It is remarkable that this corresponds to Leibniz's view of perception as isomorphic representation of the external in the internal, or of the perceived in the perceiver.[9] I use these two images to draw some similarities and, especially, contrasts between Leibniz's and Pascal's employments of infinity to depict the nature of living beings. In spite of superficial similarities, Leibniz's use of infinity to define living beings stands in stark contrast to Pascal's use of infinity in that it stresses unity and harmony, while Pascal stresses divisibility and disparity.

I will also account for the late appearance of this text by considering the development of Leibniz's views on infinity and the role it plays in his definition of living beings in the "New System" (as discussed in chapters 5 and 6). I will argue that Leibniz's reproach to Pascal that he does not make sufficient use of infinity in his description of living beings only makes sense after Leibniz's development of the notion of a natural machine presented in the previous chapter. The text will also allow us to further explore Leibniz's use of different degrees of infinity, and especially the application of infinity to created, living beings.

In sections 7.2–7.5, I present the text and the major differences between Pascal's and Leibniz's use of infinity in describing the nature of living things. In section 7.6, I offer an account of the content of this text and its appearance around 1696 by referring to the role Leibniz's view of infinity plays in his definition of living beings in the "New System." In section 7.7, I argue that, in spite of superficial similarities, Leibniz's use of infinity to define living beings stands in stark contrast to Pascal's use of infinity. Whereas Pascal uses infinity to emphasize divisibility and disparity, as well as our inability to comprehend the infinite world surrounding us, Leibniz uses infinity to emphasize the intrinsic unity each living being must have and the inherent harmony among all living beings, as well as our sense of belonging to an infinite world precisely because we, being an imitation of an *absolutely* infinite being, are infinite, too (though only in a particular way).

7.2 LEIBNIZ AND PASCAL

Throughout his life, Leibniz maintained a keen interest in Pascal's work. The evidence collected over the past century by scholars such as Baruzi,[10] Grua, and Mesnard,[11]

---

[9] In a letter to Wagner, Leibniz writes that perception consists in the "correspondence of the internal and the external, or a representation of the external in the internal, of the composite in the simple, of multiplicity in unity" (June 4, 1710; GP VII:529).
[10] Baruzi (1907); and J. Baruzi, *Leibniz* (Paris: Bloud, 1909).
[11] J. Mesnard, "Leibniz et les papiers de Pascal," *Studia Leibnitiana Supplementa* 17 (1978): 45–58.

and recently presented by Frédéric de Buzon and Maria Rosa Antognazza, clearly shows that, from early in his career, Leibniz was very well informed about Pascal's work.[12] For example, in 1671 (just a year after its publication), Leibniz bought a copy of Pascal's *Pensées*.[13] In a letter to Graevious of 1671, he speaks of Pascal's *Pensées* as a little book of gold (*libellum aureolum*), which "by the profoundness of its thought and the elegance of explication compares with any of the greatest men" (A 2.1:193). Before his arrival in Paris in 1672, and certainly during his stay there until 1676, Leibniz was in contact with the Jansenist circle (including Arnauld, Nicole, Saint Amour, Roannez, and Gilberte Pascal) and was also associated with a group loyal to Pascal ("les pascalins," as Mesnard calls them). In 1673, Leibniz conducted a study of Pascal's *Letters to A. Detonville*. Pascal's mathematical work, referred to him by Huygens, was one of the most important sources for his mathematical research from 1673 to 1676, leading to his early development of the calculus.[14]

In the beginning of 1673, Leibniz was busy developing a calculating machine expressly designed to supersede Pascal's calculating machine in performing automatic multiplication, division, and extraction of square and cube roots, in addition to summation and subtraction.[15] In 1675, Leibniz received Pascal's unedited manuscripts from E. Périer, which he studied with Tschirnhaus, and then recommended for publication in 1676.[16] This marks the last couple of years of Leibniz's stay in Paris (1675–76) as a particularly intense period in his reception of Pascal's work.[17]

Pascal's work continued to play a subtle and complex role in Leibniz's thought. Among other things, the figure of Pascal certainly was, for Leibniz, a source of inspiration, as well as a source for comparison and perhaps even of competition. If Leibniz was clearly impressed with Pascal's mathematical and experimental work,[18] his reception of his philosophical work and his methodological remarks was much more nuanced and critical.[19] The extent to which Pascal's work inspired and influenced

---

[12] F. de Buzon, "Que Lire Dans Les Deux Infinis ? Remarques Sur Une Lecture Leibnizienne," *Les Études Philosophiques* 4, no. 1 (2010a): 535–48; Antognazza (2009, 537). For more details of Leibniz's early reception of Pascal, see Mesnard (1978, 45–58).

[13] See A 1.1:436 for the receipt of Leibniz's purchase of Pascal's work.

[14] Antognazza (2009, 157–59).

[15] See A 6.2:332; see also Antognazza (2009, 144).

[16] See Leibniz's letter to Oldenburg in A 3.1:255–56. See also Antognazza (2009, 162); Buzon (2010a, 537).

[17] Mesnard even thinks that it was due to his engagement with Pascal's work that these years (1675–76) were so exceptionally rich for Leibniz. See Mesnard (1978, 58).

[18] Buzon (2010a, 538).

[19] For example, Leibniz was critical of Pascal's *Esprit géométrique* and his theory of definition. See, e.g., A 6.4:591 and 970; see Buzon (2010a, 539); and M. Lærke, *Les Lumières de Leibniz: Controverses avec Huet, Bayle, Regis, et More* (Paris: Classiques Garnier, 2015), 107–10,

Leibniz is not our present concern.[20] The present chapter focuses on a specific text—a rather brief comment Leibniz makes on fragment 22 of Pascal's *Pensées* in the Port Royal edition, 1670, then entitled *Connaissance générale de l'homme*, and including the fragment on the two infinities (*deux infinis*). However brief, this comment is of great interest—both philosophical and historical.

In this text, Leibniz refers to Pascal's notions of the infinitely large and infinitely small, and to the way Pascal uses infinity to describe living beings through the example of a mite (*ciron*). In his comment, Leibniz argues that Pascal did not go far enough in employing infinity and, in contrast to Pascal's mite, he employs an utterly different image—that of a living mirror (*miroir vivant*)—as an illustration of a living being. I compare these powerful images and draw some similarities and contrasts between Leibniz's and Pascal's employment of infinity in capturing some essential features of living beings through their respective use of these images.

Yet to be published in English (in print), this text has been the object of some studies and commentaries in German and especially in French (more details in the next section).[21] It has been recently revisited, reedited, and published by Frédéric de Buzon, with an annex presenting a version of the text that was not previously edited.[22] For the following reasons, it seems pretty clear that Leibniz composed this text sometimes around 1696: the reference to his "system of pre-established harmony, which has just recently appeared on the scene" marks the text to shortly after the "New System" (1695); the text contains the word "monad," which appears in Leibniz's text in this period, as well as the phrase "a living mirror" (*miroir vivant*). While the figure of a mirror in general appears in earlier texts, such as the "Discourse on Metaphysics," as well as in Leibniz's Paris Notes, the term *miroir vivant* shows up in this note, in his correspondence with Sophie (1696) and in his correspondence with De Volder, as well as in later texts such as the *Monadologie* 56 and the "Principles of Nature and Grace" §3, among others.[23] I am unaware of earlier occurrences of this phrase.

---

for more details. For Leibniz's attitude regarding the use of mathematics in the service of theology in relation to Pascal, see Baruzi (1907, 222–25).

[20] See, for example, Leibniz's letter to Burnett, February 1697; GP III:195; and his letter to Seckendorff, June 11, 1683; A 2.1:840. For a more thorough discussion of the way Leibniz read Pascal, see Carraud (1986, 107–24); Buzon (2010a); and Lærke (2015).

[21] See P. Riley, *Leibniz' Universal Jurisprudence: Justice as the Charity of the Wise* (Cambridge, MA: Harvard University Press, 1996), 62–64. This text is translated in Lloyd Strickland's website of Leibniz's short texts; see http://www.leibniz-translations.com/pascal.htm.

[22] Buzon (2010b).

[23] See Leibniz's Letter to De Volder; AG 177; and Leibniz to Rémond, February 11, 1715; GP III:636.

The appearance of this text circa 1696, however, presents something of a puzzle: if Leibniz knew Pascal's *Pensées* well since 1671, why did he compose *this* reaction to Pascal only twenty-five years later? Is there anything in Leibniz's development that could account for *this* text at this time, given that he had been commenting on Pascal's work throughout his career? More specifically, what prompts him to see Pascal's remarks on infinity as the entry point (*une entrée*) into his philosophical system at this point in time?[24] I will address this question toward the end of this chapter, on the basis of the development of Leibniz's thought set out in chapter 5.

Even though this text is short, it is dense, extremely rich, and interesting. One commentator has gone as far as to say that the encounter with Pascal gave Leibniz the occasion to succinctly state "toute sa philosophie."[25] Even if this is obviously overstated, there is some truth to the remark. The text is indeed one of the most succinct, condensed—and, I would say, beautiful—expressions of Leibniz's philosophy at the point his monadological phase begins to take shape.[26] In addition, written shortly after the "New System" and Leibniz's distinction between natural and artificial machines, it sheds some further light on his view of living beings at the time.

7.3 THE TEXT

As already noted, sometimes around 1696, Leibniz was busy copying fragment 22 of the so-called Port Royal edition of Pascal's *Pensées*. Once he was done with what looks like a hasty (and imprecise) copying,[27] Leibniz turns to compose a comment. His comment begins with a dramatic and curious statement:

---

[24] See Buzon (2010b): "Ce que Mons. Pascal dit de la double infinité, qui nous environne en augmentant et en diminuant, lorsque dans ses Pensées (n. 22) il parle de la connaissance générale de l'homme, n'est qu'une entrée dans mon système " (554). Buzon (2010a) articulates the question as follows: "[C]omment une sorte de résonance philosophique se met en place dans Double infinité et Monade, par laquelle un auteur peut faire entendre une pensée dans la sienne, à titre de commencement ou d'entrée, à uncertain moment de son propre développement?" (536).
[25] "On voit comment les pages de Pascal sont, pour Leibniz, l'occasion de donner, en un raccourci saisissant, toute sa philosophie" (Naërt 1985, 51).
[26] For two remarkable accounts of Leibniz's development in this period, see Fichant (2004, 7–147); and Garber (2009).
[27] In copying Pascal's text, Leibniz makes some alterations and additions. In particular, after "des cirons dans lesquels il retrouvera ce que les premiers ont donné, trouvant encore dans les autres la même chose," Leibniz adds between parenthesis, "ou des choses analogues" (second line). This suggests that Leibniz takes the *ciron* more generally as an illustration of living things. This addition by Leibniz has been noted by both Baruzi (1907, 224) and Naërt (1985, 45).

> What M. Pascal says of the double infinity, which surrounds us in increasing and decreasing, when he speaks of the general knowledge of man in his Pensées (n. 22), is but the entry point into my system.[28]

In 2010, de Buzon published a commentary in which he notes the significance of Leibniz's notion of natural machine vis-à-vis Pascal, as well as a new edition that presents two different versions of the text in meticulous detail. The first version is a marginal comment added to a transcription of the passage from the *Pensées*, with the note, "What I've added in the margin, I'd better write up in a separate paper."[29] The second version is an expansion of the marginal note, now on a separate piece of paper. While de Buzon emphasizes the similarity between Leibniz's notion of a natural machine and Pascal's view, I argue that there are significant differences in their views, which are vividly expressed in the images they use.

De Buzon's publication is the main source for the English translation presented here. Here is the first version of the text.

> Up to this point, Pascal. What he has just said of the double infinity is nothing but an entry point to my system. What wouldn't he have said with his powerful eloquence, if he had gone further, if he had known that all matter is organic, and that the least portion contains, through the actual infinity of its parts, in an infinity of ways, a living mirror expressing the entire infinite universe in an infinity of ways, such that one could read in it (if one had sufficiently penetrating sight and mind) not only the present stretched out to infinity but also the past, and all of the infinitely infinite future, since it is infinite in each moment, and since there is an infinity of moments in each part of time, and more infinity than one could ever say in all of future eternity? But the pre-established harmony goes beyond all this and yields this same universal infinity in each primary near-nothing [*presque néant*] (which is at the same time the final almost-everything [*presque tout*] and the sole thing, after God, that deserves to be called a substance), that is, [it yields this universal infinity] in each real point, which constitutes a Monad, of which I myself am one, and will not perish any more than will God or the universe, which it must always represent, being simultaneously less than God and more than the universe of matter: a diminutive God-like-thing and

---

[28] "Ce que Mons. Pascal dit de la double infinité, qui nous environne en augmentant et en diminuant, lorsque dans ses Pensées (n. 22) il parle de la connaissance générale de l'homme, n'est qu'une entrée dans mon système" (version 2, folio 213 r–v) in Buzon (2010b, 554).

[29] "Was am Rande von mir addiert, habe ich besser auf ein ander Papier geschrieben."

an eminently universe-like-thing, and as a prototype, the intelligible world being in ectype the sources of the sensible world in the ideas of God.[30]

This is obviously a dense and very complex text. It contains several astounding claims. First, Leibniz complains that Pascal does not realize that all matter is organic. This indicates that Leibniz presupposes here his panorganic view that all beings are ultimately living beings.[31] Second, organic matter, he claims, is actually divided to infinity, which, as we have seen, is a familiar theme in Leibniz. Third, and perhaps most remarkable, is the claim that, however small, each portion of matter contains a *living* mirror that expresses the infinitely large universe. Fourth, such a living mirror contains "not only the present stretched out to infinity but also the past, and all of the infinitely infinite future," which is reminiscent of Leibniz's doctrine of marks and traces that Leibniz famously ascribed to individual substances in the "Discourse on Metaphysics" (arts. 8 and 13) and elsewhere. Fifth, Leibniz's new system of preestablished harmony goes beyond all that in showing that such a living mirror captures universal infinity: being almost nothing but at the same time almost everything, it is the real point that constitutes a monad, which (sixth) is the only thing that deserves to be called a substance, after God. Seventh, Leibniz himself is such a monad (and, indeed, so are we all). Eighth, the monad is considered a diminutive God-like

---

[30] "Jusqu'ici M. Pascal. 'Was am Rande von mir addiert, habe ich besser auf ein ander Papier geschrieben.' Ce qu'il vient de dire de la double infinité n'est qu'une entrée dans mon système. Que n'aurait-il pas dit, avec cette force d'éloquence qu'il possédait, s'il y était venu plus avant, s'il avait su que toute la matière est organique, et que la moindre portion contient, par l'infinité actuelle de ses parties, d'une infinité de façons, un miroir vivant exprimant tout l'univers infini, de sorte qu'on y pourrait lire (si on avait la vue assez perçante aussi bien que l'esprit) non seulement le présent étendu à l'infini, mais encor le passé, et tout l'avenir [infini pour chaque moment] infiniment infini, puisqu'il est infini par chaque moment, et qu'il y a une infinité de moments dans chaque partie du temps, et plus d'infinité qu'on ne saurait dire dans toute l'éternité future. Mais l'harmonie préétablie passe encore tout cela et donne cette même infinité universelle dans chaque [presque néant] 'premier presque néant (qui est en même temps le dernier presque tout et le seul pourtant qui mérite d'être appelé une substance après Dieu)' c'est-à-dire dans chaque point réel, qui fait une Monade, dont moi j'en suis une, et ne périra non plus que Dieu et l'univers, qu'il doit toujours représenter, étant [un Dieu] [comme Dieu] en même temps moins qu'un Dieu et plus qu'un univers de matière: un comme-Dieu diminutif, et un comme-univers éminemment, et comme prototype, les mondes intelligibles étant en ectype les sources du monde sensible dans les idées de Dieu" (Buzon 2010b, 554).

[31] For example, the division to infinity of organic bodies and the existence of microscopic animals come up in a letter Leibniz writes to Malebranche in 1679: "There is even room to fear that there are no elements at all, everything being effectively divided to infinity in organic bodies. For if these microscopic animals are in turn composed of animals or plants or other heterogeneous bodies, and so on to infinity, it is apparent, that there would not be any elements" (A 1.2:719, translated by J. E. H. Smith, *Divine Machines: Leibniz's Philosophy of Biology* [Princeton, NJ: Princeton University Press, 2011], 160).

thing, and thus (ninth) will not perish. Finally (tenth), the monad will always represent (literally, re-present) both God and the universe.

Surely my breaking down this dense text into a list of claims can be contested. But what cannot be contested, I think, is that Leibniz manages here to bring together some of his familiar theses with some new ones in one or two remarkable sentences. Since this is one of the early appearances of the terms "monad" and "living mirror," it is not obvious how to interpret these notions here. It is fairly clear, however, that, in this passage, a living mirror is likened to a substance, and that it makes a monad, which is both active and representative; and that is exemplified through the I. The I, the Ego, or *Moy*, are recurrent examples of the true unity of substance that Leibniz uses in many other texts, both earlier and later than this one. I suspect that this point is very significant. It suggests that by "living mirror" (as well as by "monad"), Leibniz intends to refer to a complete substance, rather than to some constituent of it.[32]

Our text raises other interesting questions. For example, what is the status of the term "monad" at this point, and how does it compare with the later usage of "monad" in texts such as the *Monadologie* and the "Principles of Nature and Grace"? What does the qualification of a mirror as *living* add to the figure of a mirror, which was Leibniz already made use of in earlier texts? I will touch on these questions at the end of the chapter. In the meantime, I will mainly focus on the following theme. Like Pascal, Leibniz conceives of human beings as placed between two infinities. But, unlike Pascal, for Leibniz, human beings (as well as other living beings) are themselves seen as *infinite* creatures; and, as such, they are placed between the absolute infinity of God and the infinitely divisible matter. As we shall see, it turns out that Leibniz's use of infinity is quite different from Pascal's. Whereas, for Pascal, humans are seen as finite creatures facing and realizing their place in between the infinitely vast and the finitely minute, for Leibniz, humans are themselves placed in a graded hierarchy of infinity and perfection—"the sole thing, after God, that deserves to be called a substance ... simultaneously less than God and more than the universe of matter: a diminutive God-like thing and an eminently universe-like thing."

7.4 APPROACHING INFINITY: LEIBNIZ VS. PASCAL

Before developing this theme, I would like to bring out some of the general differences between Pascal's and Leibniz's approaches to infinity. Pascal's approach is clearly expressed in the passage to which Leibniz is responding. According to Pascal, the point of philosophical reflection is to make us realize the particularity of the human situation. Philosophical reflection reveals that we occupy a middle position between

---

[32] For further discussion of the implications of this point, see chapter 9, this volume.

two infinities: on the one hand, a universe that extends to infinity; and on the other, everything made of matter, divisible *ad infinitum*. In his *Pensées*, Pascal urges us to recognize our intermediate position between the infinitely large and the infinitely small, both of which we do not fully understand. As Pascal puts it, we know there is infinity but we do not understand its nature.[33] According to Pascal, this would lead to a realization of our true condition as finite creatures: creatures with a limited understanding facing the infinitude of the universe, as well as the infinite and incomprehensible nature of its Creator. These considerations serve as a reminder of humility in pressing the limited capacities of human reason in contrast to the infinite nature of divine wisdom and power.[34]

Pascal uses the mathematical (quantitative) sense of infinity to draw an analogy with the infinite wisdom and power of God. As he writes,

> [a] unit added to infinity does not augment it at all, nor more than does one foot add to an infinite measure. The finite annuls itself in the presence of the infinite and becomes pure nothing. So is our spirit in front of God's; so is our justice in front of divine justice.... We know that there is an infinite but we are ignorant of its nature. Just as we know that it is false that numbers are finite, so it is true that there is an infinite in number. But we don't know what it is: it is false that it would be even, it is false that it would be odd; for, adding the unit does not change its nature; regardless, it is a number and each number is either even or odd.... Thus one can well know that there is a God without knowing what it is.[35]

Pascal's aim in drawing this analogy between arithmetical infinity and the infinity of God is clear. His epistemological point, regarding the unbridgeable gap between the finite and the infinite, serves a theological purpose. The upshot of Pascal's analogy is to cast the human relation to God as a relation between a finite/limited mind and

---

[33] Fragments 233–418, in B. Pascal, *Pensées*, in *Œuvres Complètes*, ed. L. Lafuma (Paris: Éditions du Seuil, 1963), 550; see also 418: "Nous connaissons qu'il y a un infini, et ignorons sa nature."
[34] "La dernière démarche de la raison est de reconnaître qu'il y a une infinité de choses qui la surpassent. Elle n'est que faible si elle ne va pas jusque là" (Pascal 1963, 524).
[35] "L'unité jointe à l'infini ne l'augmente de rien, non plus qu'un pied a une mesure infinie. Le fini s'anéantit en présence de l'infini, et devient un pur néant. Ainsi notre esprit devant Dieu; ainsi notre justice devant la justice divine.... Nous connaissons qu'il y a un infini et ignorons sa nature. Comme nous savons qu'il est faux que les nombres soient finis, donc il est vrai qu'il y a un infini en nombre. Mais nous ne savons ce qu'il est: il est faux qu'il soit pair, il est faux qu'il soit impair; car, en ajoutant l'unité, il ne change point de nature; cependant, c'est un nombre et tout nombre est pair ou impair.... Ainsi on peut bien connaître qu'il y a un Dieu sans savoir ce qu'il est" (Pascal 1963, 550).

an infinite/unlimited one—with respect to power and with respect to knowledge and wisdom. At the same time, the relation between our finite mind and God is rather subtle. We know and recognize the infinite, but we do not understand its nature, just as we must admit infinity of number though we cannot comprehend its nature. For Pascal, the point of contemplating the infinite is precisely to make us realize our finitude and limited nature in face of the infinite and incomprehensible nature of God. Thus, according to Pascal, just as we know there is the infinite, with unambiguous clarity, even if we cannot comprehend it, so is our relation to God: we know there is a God, but at the same time we recognize that we cannot comprehend his nature. As Pascal famously writes, "we know the truth not only through reason but also through the heart" (fragments 282, 110). Unlike Leibniz, who demands a proof for both the existence of God, seen as an infinite being, and the impossibility of an infinite number, for Pascal, a clear and acute perception is all one needs and all one can ask for.

Indeed, Leibniz's attitude toward the infinite is very different. While Leibniz is acutely aware of the paradoxes threatening infinity (and especially its quantitative variants, as we have seen in chapter 2), he does not seem to be too concerned about using infinity in his philosophy.[36] Even if Leibniz is strongly committed to the impossibility of an infinite number, he is not quite optimistic about the human capacity to handle infinity. Indeed, while Leibniz argues that there is no infinite number, since this is a contradictory notion, infinity nonetheless figures in almost every aspect of his philosophy.

As we observed in the introduction, Leibniz's confidence in using infinity is most likely related to the success of his early mathematical work on infinite series, the development of the calculus,[37] and his syncategorematic interpretation of the infinitely small.[38] It goes without saying that Leibniz's method to handle the infinitely small demonstrates that a finite mind is capable of handling the infinite in this context. But

---

[36] "Among numbers there are infinite roots, infinite squares, infinite cubes. Moreover, there are as many square numbers as there are numbers in the universe. Which is impossible. Hence it follows either that in the infinite the whole is not greater then the part, which is the opinion of Galileo and Gregory of St. Vincent, and which I cannot accept; or that infinity itself is nothing, i.e. that it is not one and not a whole" (Notes on Galileo's Two New Sciences, Fall 1672; A 6.3:168; LLC 9).

[37] Ironically, it owes some of its inspiration to Pascal's mathematical work. See, for instance, Antognazza (2009, 157–59).

[38] For more discussion regarding Leibniz's approach to the infinitely small and infinitely large, see Arthur's introduction to LLC.

it is worth noting that this attitude goes against the warnings of both Descartes and Pascal of the dangers of using infinity in non-theological contexts.[39]

Given this background, it is not surprising that, in his remarks on Pascal (circa 1696), Leibniz does not criticize Pascal for using infinity. Rather, he complains that Pascal does not use infinity enough, that he has not gone far enough in describing nature as infinite, that he does not recognize how much more pervasive infinity is throughout nature, and how central it is for understanding the nature of living things and of reality itself. Hence, Pascal's reflections, Leibniz says, are but an *entrée* to his own system (*n'est qu'une entrée dans mon système*). Thus, according to Leibniz, Pascal is not so much wrong as insufficiently advanced in his employment and analysis of infinity. As we shall see in section 7.7, this subtle critique implies some significant differences between Leibniz and Pascal.

7.5 PASCAL'S MITE AND LEIBNIZ'S LIVING MIRROR

Leibniz's general reproach in this text seems fairly clear. While Pascal went a long way in ascribing infinity to nature, he did not go far enough. However, since Pascal fully embraces infinity, Leibniz's reproach is not so clear, after all; if Leibniz's reproach is justified, it must be more specific. Here is what Leibniz says in the second version of the text:

> What wouldn't he have said, with his powerful eloquence, if he had gone further, if he had known that all matter is organic throughout, and that however small a portion one takes contains, representatively, by virtue of the actual diminution to infinity that it encompasses, the actual increase to infinity that is in the universe outside that portion—that is to say, that each little portion contains, in an infinity of ways, a living mirror expressing the entire, infinite universe?[40]

Leibniz argues here that, however small, each part of matter is organic and makes up a living mirror that expresses the whole universe. His use of this image of the living mirror is both new and curious. While he refers to the notion of a mirror much earlier in his career, to the best of my knowledge, this is the first time that the notion of a *living* mirror appears in his

---

[39] In fact, Leibniz's method consists in translating infinite magnitudes into finite ones, just smaller than any assignable, but the details of this method go beyond our immediate concerns here.

[40] Buzon (2010b, 554).

writings.⁴¹ Leibniz's use of this curious image draws on Pascal's reference to both the infinitely small and the infinitely large, but alters it such that each portion of matter, however small, *represents* the infinitely large universe. For Leibniz, a living mirror, which may be smaller than any given size, expresses a universe that may be larger than any assignable magnitude.⁴²

An important point, to which I shall return, is that such a mirror is an active living being—*c'est un miroir vivant,* he stresses—such that the mirroring is related to the inner activity (of perception) rather than passive reflection common in a regular mirror. Recall again that, according to Leibniz, the ontological bedrock of the actual world consists of organic things. As he writes, "all organic bodies are animate, and all bodies are either organic or collections of organic bodies" (A 6.4:1798; LLC 277). As we know from later texts, from this period on, the Leibnizian universe becomes populated with infinitely many living beings for which he will adopt the term "monads."⁴³ In light of his view that nature consists of living beings and that an essential feature of living beings is that, however small, their inner nature represents (while being a part of) the infinitely large universe, Leibniz's response to Pascal seems not only more specific but also to signal a radical break from the way Pascal uses infinity, as well as from his interpretation of both the infinitely small and the infinitely large. For, in Leibniz, each infinitely small living mirror mirrors (in the sense of expressing, representing) the infinitely large universe.

Let us compare Pascal's depiction of a mite with Leibniz's depiction of a living being as a living mirror. Here is what Pascal says:

> What is a man in the infinite? Who can comprehend it? But to show him another prodigy equally astonishing, let him examine the most delicate things he knows. Let a mite be given him, with its minute body and parts incomparably more minute, limbs with their joints, veins in the limbs, blood in the veins, humours in the blood, drops in the humours, vapours in the drops. Dividing these last things again, let him exhaust his powers and his conceptions, and let the last object at which he can arrive be now that of our discourse. Perhaps he will think that here is the smallest point in nature. I will let him see therein a new abyss. I will paint for him not only the visible universe, but also everything he is capable of conceiving of nature's

---

⁴¹ For a survey and some analysis of Leibniz's use of mirrors, see C. Marras, "Mirrors That Mirror Each Other," in *Papers from the VIII Internationaler Leibniz Kongress,* ed. Herbest Breger, Jürgen Herbest, and Sven Erdner (Hannover: n.p., 2006), 556–64.
⁴² With much insight, but without any explication, Baruzi remarks, "[a]insi se transforme le 'ciron' de Pascal" (Baruzi 1907, 227).
⁴³ See, for example, *Monadologie* 56, 83; PNG §3; AG 207–208.

immensity in the womb of this imperceptible atom. Let him see therein an infinity of worlds, each of which has its firmament, its planets, its earth, in the same proportion as in the visible world; in this earth of animals, and ultimately of mites, in which he will find again all that the first had, finding still in these others the same thing without end and without cessation. Let him lose himself in wonders as amazing in their minuteness as the others in their vastness.[44]

Pascal's imagery here is quite astonishing. In fact, it seems remarkably similar to Leibniz's early view that the infinitely small implies new *abimes* in the form of worlds within worlds to infinity.[45] But there is an important difference: in Leibniz's notion of living mirror, the nested structure that develops to infinity *represents and mirrors* the infinitely vast world by virtue of active perception. For Leibniz, there is an inherent, structural connection between the infinitely small and the infinitely large in the very constitution of the world. The two infinities are not disparate, as in Pascal but, rather, intrinsically connected, in the sense that they map onto one another.[46] In this way, each minute constituent of the world expresses all the rest through isomorphic

---

[44] The passages continues thus: "For who will not be astounded at the fact that our body, which a little while ago was imperceptible in the universe, itself imperceptible in the bosom of the whole, is now a colossus, a world, or rather a whole, in respect of the final smallness which we cannot reach? He who regards himself in this light will be afraid of himself, and observing himself suspended in the mass given him by nature between those two abysses of the Infinite and Nothing, of which he is equally removed, will tremble at the sight of these marvels; and I think that, as his curiosity changes into admiration, he will be more disposed to contemplate them in silence than to examine them with presumption." Lafuma §199. For the English cited here, see G. W. Leibniz, "Double Infinity in Pascal and Monad," trans. Lloyd Strickland, *Leibniz-Translations.com*, n.d., http://www.leibniz-translations.com/pascal.htm.

[45] As early as his "Theory of Concrete Motion" (1670–71), Leibniz articulates the doctrine (mentioned by Pascal as well) that, in every bit of matter, there are worlds within worlds, and that this goes on to infinity. In this context, the doctrine appears as a consequence of the infinite divisibility of the continuum. Leibniz writes: "any atom will be of infinite species, like a sort of world, and there will be worlds within worlds to infinity" (A 6.2 N40; LLC 338–39). A similar view appears several years later in Leibniz's notes from Paris (1676), where he writes that every part of the world, regardless of how small, "contains an infinity of creatures," which is itself a kind of "world" (A 6.3:474). Similarly, in the dialogue "Pacidius to Philalethes," Leibniz says, "in any grain of sand whatever there is not just a world, but an infinity of worlds" (A 6.3:566; LLC 211).

[46] This point should be qualified, however. In his *L'esprit de la Géométrie*, art. 19, Pascal writes, "the two infinities, while infinitely different are still related [*relatifs*] to one another, so that knowledge of the one necessarily leads to knowledge of the other" (in B. Pascal, *"L'esprit de la Géométrie*, ed. B. Clerté and M. Lhoste-Navarre [Paris: Bordas, 1986], 34, my translation). The relation Pascal notes, however, is clearly epistemic, whereas Leibniz emphasizes an ontological connection as well.

relations, which are at the heart of his system of pre-established harmony.[47] Leibniz's notion of living mirror is thus consistent with the homomorphism he sees between each constituent of the world and the world as a whole. "C'est tout comme ici, partout et toujours," as we observed in the previous chapter.[48] The inner structure of each monad resembles the structure of any other, so that active perception of its inner activity would mirror that of all other monads and, thus, of the world (consisting of these monads). The mirroring is active as each living being perceives its own infinite structure, which represents that of the whole world.

It is also worth stressing that Leibniz's view here coheres with his views on perception. Indeed, it is part and parcel of his view of perception as the characteristic activity (besides appetition) of monads. What is distinctive about *perceptual* or *mental* representation is that, in it, "many things are expressed in one" (PNG §2; *Moodologie* 14). In a letter to Wagner, Leibniz writes that perception consists in the "correspondence of the internal and the external, or a representation of the external in the internal, of the composite in the simple, of multiplicity in unity" (June 4, 1710; GP VII:529). In the PNG, Leibniz notes that perception is,

> the representation of the composite or what is external, in the simple. . . . For the simplicity of substance does not prevent a multiplicity of modifications, which must be found together in this same simple substance, which must consist of the variety of its relations to external things. Similarly, in a center or point, though entirely simple, we find an infinity of angles formed by the lines that meet there. (PNG 2; AG 207)

The explicit connection with infinity is obviously pertinent here. A single and infinitely small unit may reflect an infinity of others, as a point at the center of a circle may reflect infinitely many angles. Commentators have noted that, for Leibniz, perception is strongly related to expression—that it is a special kind of representation, an expression in the sense of a certain isomorphism or structural similarity between the internal state of the perceiver and the external state (See Kulstad 1977; Simmons 2001; Jorati 2017). But what Leibniz's comment on Pascal suggests is that this isomorphic

---

[47] In order to avoid any misunderstandings, it is perhaps worth emphasizing that Leibniz's view does not imply the existence of infinitely small beings, which Leibniz flatly denies. For Leibniz, infinitesimals are not entities, but useful fictions. Rather, for any finite living being, no matter how small, there is a smaller one. This is what the actual infinity involves, according to Leibniz.

[48] See Leibniz to Sophie Charlotte, May 8, 1704; GP III:343–48. For this reason, "God sees in each portion of the universe the whole things. . . . He is infinitely more discerning than Pythogoras, who judged the height of Hercules by the size of his foot point" (Theodicy §341).

relation is also mediated through the active perception of the monad's internal complexity. I will get back to the internal complexity of the monad in chapter 9.

7.6 THE WONDERS OF INFINITY AND THEIR THEOLOGICAL UNDERTONES: DESPERATION VS. CELEBRATION

As we have just seen, there is a sharp difference in deploying infinity between Pascal and Leibniz. For one thing, there is a strategic difference in what they use infinity for. The aim of Pascal's reflection on the infinite is that awareness of its paradoxical nature would reveal our true nature as finite, cognitively and rationally limited beings, and thus incapable of comprehending the infinity of nature surrounding us. In particular, Pascal's aim in his description of a mite is to astound and even shock his readers. Obviously, Pascal's ultimate goal here is not a detached, "scientific" description of the living world.[49] Rather, he urges his readers to lose themselves (*qu'il se regarde comme égaré*) in the marvels of infinity, which are as astonishing in their vastness as in their minuteness. But these marvels are meant to bring out the inherent disproportion and frustration of the human intellect, as it finds itself caught up in between these infinities, "incapable of understating the infinity which surrounds us."[50]

Leibniz is obviously impressed with the wonders of infinity, as well. But he is not as impressed with the idea of losing ourselves in its wonders. Whereas, for Pascal, the human mind loses itself in despair between the infinitely large and the infinitely small, being lost or in despair are both foreign to Leibniz's spirit. Instead, he is well known for his optimism—though the celebration of the infinite is one aspect of his optimism that has not been sufficiently appreciated.[51] In contrast with Pascal's

---

[49] For an elaboration of Pascal's attitude, stressing the consideration of human beings rather than the contemplation of nature, see Carraud (1992, 403–34, secs. 30–31). Carraud also adds the following perceptive remark: "Le regard pascalien est regard sur l'autre aveugle, regard sans être regardé, sans réciprocité, sans miroir" (397). He notes that the notion of a mirror, so typical of the Renaissance, does not appear even once in the *Pensées*. It is all the more striking, therefore, that Leibniz is contrasting the notion of living mirror to that of Pascal's mite.

[50] "Car enfin qu'estce que l'homme dans la nature? Un néant à l'égard de l'infini, un tout à l'égard du néant, un milieu entre rien et tout, infiniment éloigné de comprendre les extrêmes. La fin des choses et leurs principes sont pour lui invinciblement cachés dans un secret impénétrable, également incapable de voir le néant d'où il est tiré et l'infini où il est englouti" (Lafuma §199, 526); see also http://www.penseesdepascal.fr/Transition/Transition4-moderne.php.

[51] This is especially the case when contrasted with the notorious aspect of his optimism made infamous by Voltaire in *Candide*. For contrasting Leibniz's optimistic attitude with Pascal's pessimistic one, though in a different context, see Naërt (1985, 167), and in yet another context, see Lærke's recent perceptive remark: "La situation du géomètre leibnizien se rapproche beaucoup de celle de l'"homme, dans l'infini' dont parle Pascal, et qui se trouve 'suspendu

marvel and *desperation* in face of the infinite,[52] in Leibniz we find marvel and *celebration* of the infinite instead. As noted earlier, infinity figures in almost every aspect of Leibniz's philosophy. This is vivid in our text: "the least portion contains, through the actual infinity of its parts, in an infinity of ways, a living mirror expressing the entire infinite universe in an infinity of ways, such that one could read in it ... not only the present stretched out to infinity but also the past, and all of the infinitely infinite future." In this way, Leibniz turns Pascal's despairing attitude into a celebration of infinity. While Pascal attempts to make our rational aspirations more humble, Leibniz is ever optimistic about the capacity of human reason to further extend itself in general, and with respect to the notion of infinity in particular.

Like Pascal, Leibniz's attitude has some strong theological motivation, as well. In fact, both thinkers believe that the contemplation of the infinite would lead us to God.[53] But it would do so in very different ways. For Leibniz, celebrating infinity is strongly related to his conviction that infinity is an essential aspect of nature in general, and of our nature in particular. Therefore, studying infinity constitutes a way not only to appreciate and admire the glory of God as its creator but also to appreciate our deep connection to nature. Rather than despair in the labyrinthine and awesome nature of infinity and our disproportion to it, as Pascal suggests, Leibniz maintains that we should study and appreciate infinity as a constitutive and positive aspect of nature, including our own. More precisely, infinity is an important aspect of the way created things are made in the image of God. To put it differently, for Leibniz, infinity is part of the likeness between God and his creatures, so that the infinite in nature manifests that of its creator. Thus, Leibniz's extensive use of infinity in describing the natural world derives not only from his mathematical work but also from some theological and metaphysical commitments. Leibniz's mathematical work has taught him how to treat the quantitative infinite in a rational manner, but he undoubtedly uses it to further his theological ends.

Some of Leibniz's theological commitments may resonate with what Lea Schweiz has recently called "a sacramental view of nature." According to this view, "the whole of the created order [can be seen] as exhibiting one of the key principles of Lutheran sacramental theology, namely, that the finite is capable of the infinite [*finitum capax*

---

dans la masse que la nature lui a donnée entre ces deux abîmes de l'infini et du néant, dont il est également éloigné' [Lafuma §199]. 'Toutefois, et contrairement à la vision plutôt austère de Pascal, pour Leibniz, cette suspension dans l'infini n'a rien d'épistémologiquement tragique: Pascal exige trop de la science démonstrative'" (Lærke 2015, 172).

[52] "Que fera(t)il donc sinon d'apercevoir quelque apparence du milieu des choses dans un désespoir éternel de connaître ni leur principe ni leur fin" (Lafuma §199, 526).

[53] "Ces extrémités se touchent et se réunissent a force de s'être éloignées et se retrouvent en Dieu, et en Dieu seulement" (Lafuma §199, 527).

infiniti]."⁵⁴ The finite, created world is made in the image of God and, for this reason, it is seen as capable of presenting and manifesting the infinite essence and perfection of God. This theological commitment was certainly controversial. In Malebranche, for example, we find a diametrically opposed view.⁵⁵ This commitment goes some ways toward explaining why Leibniz complains that Pascal, despite having made so much of the notion of infinity, did not go far enough. It would also explain why Leibniz finds Pascal's description of the infinitely large and the infinitely small on the right track, but that it ultimately remains only the entry point to his new system of pre-established harmony.

Leibniz's ascription of infinity to the created world through the principle that the finite is capable of the infinite raises, however, a serious problem. The relation between God and his creatures is commonly understood at the time as a categorical gap between an infinite entity and finite entities. If Leibniz regards creatures as infinite as well, how would he account for the difference between creatures and the Creator? In fact, Leibniz's comment on Pascal provides us with an important clue. Most of the new philosophers—including Pascal, Descartes, and Spinoza—endorsed this dichotomy. The gist of Leibniz's approach is to cast the difference between creatures and the Creator not in terms of a categorical divide but, rather, in degrees of infinity. Leibniz's description of a living mirror in our text as "being simultaneously less than God and more than the universe of matter: a diminutive God-like thing" bears out this notion of degrees. Traditionally, the categorical distinction between finite and infinite was seen as capturing the distinction between God and individual things. In sharp contrast to this tradition, our text suggests that Leibniz draws the distinction in terms of degrees: the absolute infinity of God is set above the infinity of creatures, which is set above the infinite divisibility of matter and the infinity of mathematical

---

⁵⁴ As Schweitz writes, "Lutheran sacramental theology affirms that finite matter in the forms of bread, wine, and water is a means of grace and a vehicle for the divine. Said another way, the sacraments are instances when the 'finite is capable of the infinite.' The material elements of the sacraments are means of real and transformative encounters with the divine because they are capable of the infinite in, with, and under the finite" (L. F. Schweitz, "On the Continuity of Nature and the Uniqueness of Human Life in G. W. Leibniz," in *The Life Sciences in Early Modern Philosophy*, ed. O. Nachtomy and J. E. H. Smith [New York: Oxford University Press, 2014], 214). For further information on Leibniz's Lutheran heritage, see U. Goldenbaum, "Leibniz as a Lutheran," in *Leibniz, Mysticism, and Religion.*, ed. A. Coudert, Richard H. Popkin, and Gordon M. Weiner (Dordrecht: Kluwer Academic, 1998), 169–92.

⁵⁵ "On ne peut concevoir que quelque chose de créé puisse représenter l'infini; que l'être sans restriction, l'être immense, l'être universel puisse être aperçu par une idée, c'est a dire, par un être particulier, par un être différent de l'être universel & infini. Mais pour les êtres particuliers, il n'est pas difficile de concevoir qu'ils puissent être représentés par l'être infini qui les renferme dans sa substance très efficace, et par conséquent très intelligible" (*Recherche de la vérité*, V III, II, VII, 11; Pascal 1963, I:449).

things. This is consistent with an earlier distinction Leibniz draws between three degrees of infinity that apply to three degrees of being (see chapter 3).[56]

Our text points to the way Leibniz conceives of the relation between the finite and the infinite in contrast to the way Pascal does. It suggests that Leibniz understands the gap between the finite and the infinite not merely as a categorical distinction, as it was traditionally understood, but also as one of degree. Thus, for Leibniz, every created thing is seen as infinite; but infinity (as well as perfection) turns out to admit of different degrees (as we have seen in Leibniz's annotations on Spinoza's letter 12 in chapter 2). I will return to Leibniz's notion of degrees of infinity in the next chapter. In the next section, I turn to the question I raised at the beginning: how to account for the fact that *this* particular reaction to Pascal comes up only at this stage of Leibniz's career (circa 1696)?

7.7 WHY *THIS* RESPONSE TO PASCAL AT *THIS* TIME (CIRCA 1696)?

In the first section, Leibniz was familiar with Pascal's work and had commented on it since very early in his career. However, Leibniz drafts this comment on Pascal shortly after his "New System." What could account for his response to Pascal on the double infinity at this point in time? Perhaps some changes that took place in Leibniz's own views can account for this response at this time. Indeed, attending to the development of Leibniz's definition of life in terms of infinity can throw some light on this question. As we saw in the previous two chapters, the notion of a natural machine with its nested structure to infinity comes to the foreground as Leibniz's prime model of living beings only after 1695.[57]

In the "New System" (1695), Leibniz is no longer using infinity merely to describe nature as worlds within worlds to infinity, as he has done previously; instead, infinity now becomes one of the defining features of living beings. By virtue of being a divine creation rather than a result of human production, a natural machine is both infinite (thus presenting the mark of its creator) and a single, indestructible entity, which remains one and the same as long as it acts. Unlike an artificial machine, a natural machine cannot be composed or decomposed. It is created as one functional unit, however complex its internal states may be. As a consequence, it remains the same as long as it lives—which is forever, unless annihilated by God. Hence, a natural machine always preserves a certain degree of life or primitive activity. What gives a natural machine—a machine with an infinitely complex

---

[56] See his note from February 1676; A 6.3:385; LLC 43, cited in chapter 2, this volume.
[57] See Fichant (2003); Duchesneau (2010); Nachtomy (2010); Smith (2011); and Arthur (2014).

structure—its unity is an internal law of production. This internal law functions as a program for self-organization and self-regulation, such that each Leibnizian substance is also causally self-sufficient. According to Leibniz, a living being is infinite both in the sense of being ever active and in its nested structure, *ad infinitum*.[58] The infinity and unity of living beings are intrinsically related, of course, to the fact (noted earlier) that they are "divine machines," created by an infinite creator.[59]

It is arguable that Leibniz's view of a natural machine is very similar to Pascal's description of a mite, such that each of its parts is further divisible to infinity. Frédéric de Buzon has recently observed such similarity:

> That the parts of living beings are also living beings, and this to infinity, is exactly Leibniz's conception of natural machines, whose difference from artificial machines is only that they are "machines to the least of their parts."[60]

While de Buzon is certainly right in pointing to the notion of a natural machine as the most pertinent novelty as the context for Leibniz's comment on Pascal, and despite the similarity he observes, there is also a significant dissimilarity in the role infinity plays in Leibniz's and Pascal's respective views of living beings. Whereas for Pascal the infinitely small is related to the divisibility of matter, for Leibniz, the infinity of a natural machine is related instead to the intrinsic unity and indestructibility of substances. According to Leibniz, the distinctive feature of a natural machine (in distinction from an artificial machine) is that it is *not* infinitely divisible. In fact, it is not divisible at all. Rather, the infinite structure of a natural machine is what makes it an indivisible and indestructible unity. For the unity and indestructibility of a natural machine derives from its internal law of development—a unifying law that informs the change of its states to infinity. This law is also what makes it infinite. Indeed, infinity is also what makes a natural machine or a divine machine—that is, a machine that cannot be composed or decomposed by humans but is created as a natural unity by God.

Since Leibniz begins to articulate this conception of a natural machine in the "New System" of 1695, we can see why *this* particular response to Pascal was not likely to come up earlier in Leibniz's career despite his long familiarity with Pascal's work. Given this background, it need not surprise us that Leibniz would respond by

---

[58] See also Leibniz to Lady Masham, 1704; GP III:356, cited in chapter 6, this volume.
[59] See also Leibniz's fifth letter to Clarke, arts. 115 and 116; AG 344–45; GP VII:417–18.
[60] "Que les parties des êtres vivants soient aussi des êtres vivants, et ce à l'infini, est exactement la conception des machines de la nature, dont la différence avec les machines de l'art est que les premières sont 'machines jusques dans leurs moindres parties'" (Buzon 2010a, 547, my translation). See also *Considérations sur les principes de vie et sur les natures plastiques*; GP VI:543.

claiming that Pascal did not see the full significance of infinity as a defining feature of living beings shortly after coming to define living beings through the nested structure *ad infinitum* of natural machines.

So far so good, but one might wonder at this point what the relation is between the notion of a natural machine and that of a living mirror. The term "mirror" does not appear in the "New System." Leibniz, however, comes rather close to implicating it in several passages in which he discusses the representative nature of the soul:

> This is what makes every substance represent the whole universe exactly and in its own way, from a certain point of view. . . . And since this nature that pertains to the soul is representative of the universe in a very exact manner (though more or less distinctly), the series of representations produced by the soul will correspond naturally to the series of changes in the universe itself. . . . Since every mind is like a world apart, self-sufficient, independent of any other creature, containing infinity, and expressing the universe, it is as durable, subsistent, and absolute as the universe of creatures itself. (AG 143–45; GP IV:484–86)

The definition of a natural machine as a machine in the least of its parts implies a view of an animate being with an infinitely complex structure of machines within machines to infinity. I called this feature a nested structure that develops *ad infinitum*. Against this background, depicting a living being as a living mirror brings out a new feature of Leibniz's view explored in the two previous chapters: the inner perception of its proper structure (of the infinitely small, in Pascal's terms) allows a representation of the infinitely large, by virtue of the isomorphic relation between the inner structure of each living being and that of all others. Hence, the role of active mirroring derives from the inner perception that also represents the external world. Perhaps this is why the figure of a mirror that already appears in the "Discourse on Metaphysics" (art. 9) now becomes *living*, so that it now comes to exemplify the very nature of a living or animate being. It is worth observing that the notions of mirror and living mirror also come up in other texts circa 1696. Let us consider Leibniz's letter to Sophie, written November 4, 1696.[61]

In this letter to Sophie, we find Leibniz expressing many of the points just noted (while using the terms "mirror," "living mirror," and "machine of nature") and strongly echoing some of the doctrines presented in the "New System." After noting that some Cartesians have complained that he attempts to reestablish the view that animals are entitled to have souls (*âmes*) and that all bodies involve some vigor and

---

[61] A 1.12:90–93. I use Strickland's English translation of the letter here (Leibniz 2006).

life (*de la vigueur et la vie*), rather than being mere extended mass (in Fichant 2004, 333), he writes these famous lines:

> My fundamental meditations turn on two things, namely on unity and on infinity. Souls are unities and bodies are multitudes, but infinite ones, so that the slightest grain of dust contains a world of an infinity of creatures. And microscopes have revealed more than a million living animals in a drop of water. But unities, even though they are indivisible and without parts, nevertheless represent the multitudes, in much the same way as all the lines from the circumference are united in the centre of the circle, which alone faces it from all sides even though it does not have any size at all. The admirable nature of sentiment consists in this reunion of infinity in unity [*cette réunion de l'infini dans l'unité*], which also makes each soul like a world apart, representing the larger world in its way and according to its point of view, and that consequently each soul, once it begins to exist, must be as durable as the world itself, of which it is a perpetual mirror. These mirrors are likewise universal, and each soul exactly expresses the universe in its entirety. (Leibniz 2006, 79)

The similarity between this text and Leibniz's comment on Pascal is striking. A bit later in the text Leibniz notes that the *secte Machinale* has gone too far in reducing animals to machines, and thus downgrading the majesty of nature (repeating the very phrase he used in the "New System"). He then argues that, once we would have a better grasp of the infinite, we would have an altogether different idea of nature, in seeing its majesty rather than seeing it as reduced to mere machines, or, as nothing more than a workman's shop (*la boutique d'un ouvrier*), as the otherwise clever author of the *Entretiens sur la pluralité des mondes* (viz., Fontenelle) believes. Leibniz then continues:

> The Machines of nature are infinitely above ours. For besides the fact that they have sensation, each contains an infinity of organs, and what is even more marvelous,[62] it is for that reason that every animal is resistant to all accidents and can never be destroyed, but only changed and strengthened by death, as a snake sheds its old skin. (Leibniz 2006, 80–81)[63]

---

[62] My translation differs slightly from Strickland's (Leibniz 2006) here.

[63] "Les Machines de la nature sont infiniment au-dessus des nôtres. Car outre qu'elles ont du sentiment, chacune contient une infinité d'organes; et ce qui est encore plus merveilleux, c'est par cela que chaque animal est a l'épreuve de tous les accidents et ne saurait être jamais détruit, mais seulement change et resserré par la mort, comme un serpent quitte sa veille peau" (Fichant 2004, 336–37).

The term "living mirror" comes up at the very end of the letter:

> And it is in this that consists the advantage of minds [*esprits*] for which the sovereign Intelligence has made everything else, so as to make itself known and loved, multiplying itself so to speak in all these living mirrors that represent it. (Leibniz 2006, 81)

The resemblance between Leibniz's use of living mirrors here and his use of monads is striking. Whatever else it may signify, the term "monad" clearly indicates a genuine unity. As I have suggested, the notion of a living mirror provides us with some clues as to how a genuine unity can be made compatible with infinity—that is, its having an infinitely complex structure that resembles that of the world. This line seems to be strongly supported by Leibniz's remarks on infinity and unity in the letter to Sophie just cited.

7.8 DIVISIBILITY AND DISPARITY VS. UNITY AND HARMONY

As we have seen, for Leibniz, the infinity of natural machines is not a principle indicating division, as in Pascal; rather, it is a principle indicating unity. This also explains why the term "monad" is used in this context. The unity of a natural machine with its structure of machines nested one within the other to infinity derives from its inner source of activity—its entelechy, which is in turn informed by an internal law of development.

Furthermore, Leibniz's usage of infinity in his comment on Pascal not only signals unity as opposed to divisibility but also harmony and connectedness as opposed to the disparity and disproportion Pascal emphasizes. The notion of a living mirror not only encapsulates the infinitely small but also allows a *representation of* the infinitely large by virtue of internal perception.[64] As Leibniz writes to De Volder, a living mirror is a "concentrated world," a "world" whose inner structure expresses the structure of the universe. Further, a living mirror expresses the world through active perception, whose role is to reveal the diversity of each such individual through its active principle. This, I believe, is why Leibniz relates the notion of a living mirror to that of an entelechy:

> Entelechies must necessarily differ, that is, they must not be entirely similar to each other. Indeed, they must be sources [*principia*] of diversity, for

---

[64] For an interesting discussion of related issues, see Serres (1968, 720–22).

different ones express the universe differently, each from its own way of viewing things; it is their duty to be so many living mirrors of things, that is, so many concentrated worlds. (Leibniz to De Volder, June 1703; AG 177; GP II:251–52)[65]

Leibniz's use of infinity through the notion of a living mirror suggests that each individual being, no matter how minute, forms an integral part of a well-connected and harmonious system. Whereas Pascal exploits the infinite division of the organic world to stress our alienation and incomprehension of the world surrounding us, for Leibniz, infinity serves to bring out a sense of connectedness between individual substances, a sense of harmony, and, for that reason, one might even say, a certain sense of belonging rather than alienation.[66] Indeed, for Leibniz, infinity need not make the world strange and incomprehensible to us. Rather, being made in the image of God, we are infinite as well, and should feel at home in a world in which every aspect bears the mark of an infinite creator.[67] Indeed, as living mirrors, we not only resemble the entire infinite world but, in living, we express the infinite world, albeit from a certain (limited) perspective.

7.9 CONCLUSION

Unlike most thinkers of the period, including as Descartes, Malebranche, and Pascal, Leibniz ascribes infinity to created beings as one of their essential features. He rejects the sharp dichotomy between an infinite Creator and finite creatures, as well as the epistemological imperative (explicit in both Descartes and Pascal) that, as finite minds, we cannot, and thus should not even attempt to, grasp the infinite. By contrast, Leibniz argues that the infinite need not be dreaded but, rather, investigated, so that the glory of God and its expression in the created world would become more apparent and comprehensible. Thus, for him, created substances are imitations of their creator *in this respect* (infinity). As argued here, the kind of infinity related to being is not quantitative, so that creatures do not possess an infinitesimal *magnitude*. It is, rather, infinity related to a program of action that would last as long as creatures would act. As Leibniz writes in his note on Pascal,

---

[65] For more on the connection between entelechies and living mirrors, see GP VI:626.
[66] "L'homme, dans l'infini" se trouve "suspendu dans la masse que la nature lui a donnée entre ces deux abîmes de l'infini et du néant, dont il est également éloigné" (Buzon 2010b, 551–52; Lafuma §199).
[67] At the same time, the kind of infinity Leibniz ascribes to created beings is not the same as the absolute infinity he ascribes to God. It is also not the (quantitative) infinity he employs in mathematics. I attend to this question in the next chapter.

all these wonders are surpassed by the envelopment of what is infinitely above all greatnesses in what is infinitely below all smallnesses—that is, our pre-established harmony, which has only recently appeared on the scene, and which yields even more than absolutely universal infinity, concentrated in the more than infinitely small and absolutely singular, by placing, virtually, the whole series of the universe in each real point, which constitutes a Monad, or substantial unity, of which I am one – that is, in each truly singular, unique substance that is the fundamental [*primitif*] subject of life and action, always endowed with perception and with appetition, always containing within what it is the tendency to what it will be, in order to represent all else that will be.[68]

Leibniz goes on to say that this substantial unity, or a "living mirror," which is the fundamental subject of life and action, is a "diminutive God-like thing." It is like God in that it is a living, active being. But, unlike God, it is a particular and thus limited expression of God, whose perceptions are often indistinct and confused.

Leibniz's response to Pascal thus brings out clearly the relation he sees between infinity and living beings. On the face of it, Leibniz does not dispute Pascal's description of living beings as infinite; he argues that Pascal did not go far enough in ascribing infinity to the organic world. But, as suggested here, it turns out that Leibniz's sense of infinity and the use he makes of it are rather different from Pascal's. Had Pascal comprehended the true nature of the organic world, on Leibniz's understanding, he would see that infinity cuts deeper into the nature of things, that it is the mark of living beings that constitutes the fundamental ontology of the universe. Furthermore, each living being mirrors the whole universe by virtue of being itself infinite and thus constitutes a living representation of the universe. Leibniz's notion of living mirror illustrates his view that each living being, whose inner structure develops to infinity, actively represents the (infinitely large, in Pascal's terms) world. At the same time, it is a principle of unity that stands above the infinite divisibility of matter.

While the wonders of infinity invoke awe and astonishment, they also deserve admiration and contemplation. Leibniz is convinced that contemplating and studying the infinite will yield a sense of comprehension and belonging rather than a sense of fear, alienation, and despair.

[68] "[T]outes ces merveilles sont effacées par l'enveloppement de ce qui est 'infiniment' au-dessus de toutes les grandeurs dans ce qui est 'infiniment' au-dessous de toutes les petitesses; c'est-à-dire notre harmonie préétablie, qui vient de paraître aux hommes depuis peu, et qui donne cette même plus qu'infinité 'tout à fait' universelle, concentrée dans le plus qu'infiniment petit tout à fait singulier, en mettant virtuellement toute la suite de l'univers dans chaque point réel qui fait une Monade 'ou unité substantielle' dont moi j'en suis une; c'est-à-dire dans chaque substance véritablement une, unique, sujet primitif de la vie et action, toujours doué de perception et appétition, toujours renfermant avec ce qu'il est la tendance à ce qu'il sera" (Buzon 2010b, 555).

8

CREATED THINGS AS INFINITE AND LIMITED

8.1 INTRODUCTION

My starting point in this chapter is the characterization of God as infinite in a hypercategorematic sense—that is, a being beyond any determination. By contrast, creatures are determinate beings; they are determinate and thus limited and particular expressions of the divine essence. But since for Leibniz both God and creatures are infinite, creatures are seen as infinite and limited. This leads to seeing creatures as infinite in kind, in distinction from the absolute and hypercategorematic infinity of God. I present three lines of argument to substantiate this point: (1) seeing creatures as entailing a particular sequence of perfections and imperfections; (2) seeing creatures under the rubric of an intermediate degree of infinity and perfection that Leibniz calls maximum in kind in 1676; and (3) observing that primitive force, a defining feature of created substance, may be seen as infinite in a metaphysical sense. This leads to viewing Leibniz's use of infinity within a Neoplatonic framework of descending degrees of being, from the hypercategorematic infinite, identified with the most perfect being, to the intermediate degree of maximum in kind, identified with creatures, to the lowest degree of *entia rationis* (or beings of reason), identified with mathematical and abstract entities.

In a rare and revealing use of the term "hypercategorematic infinite," Leibniz writes to Des Bosses in 1706,

> In addition [to the syncategorematic infinite], there is a *hypercategorematic infinite*, or potestative infinite, an active power having, as it were, parts eminently but not formally or actually. This infinite is God himself. (LDB 52–53; GP II:314–15)

In a recent article,[1] Maria Rosa Antognazza provides a careful and insightful analysis of this passage. Concerning Leibniz's use of the term "hypercategorematic," she writes,

---

[1] Antognazza (2015, 5–30).

It seems clear that by hypercategorematic Leibniz means that which is beyond all categoremata, namely, that which is beyond any determinate thing falling under the Aristotelian categories and signified by categorematic terms. In other words, hypercategorematic is that which is beyond any determination. (11)

Since Leibniz tells us clearly in this passage that "This infinite is God himself," this qualification may certainly shed some light on his view of God. God is seen as an infinite being who possesses infinite, active power that goes beyond any determination—and contains all parts eminently but not formally or actually. At the same time, this qualification, being beyond all determination, also reveals something important about Leibniz's view of the relation between God and creatures. For creatures, according to Leibniz, are defined precisely as determinate and particular (and thus limited) manifestations of the divine essence. In fact, we can gain insight into Leibniz's view both of the Creator and of creatures by reflecting on this contrast: a creature is a determinate thing while God is beyond any determination—and thus may be regarded as infinite in a hypercategorematic sense. Since creatures, for Leibniz, are also infinite in some respects, perhaps we could say that creatures are infinite in a determinate way. In what follows, I try to substantiate this point.

As Antognazza insightfully notes,

> In introducing this unusual notion, Leibniz is breaking new ground. Although this expression appears to occur only in the 1706 passage for Des Bosses, once unpacked and read in conjunction with other statements about God, it sheds light on some fundamental features of Leibniz's conception of God and his relation to creatures. (11)

Indeed, part of the new ground Leibniz is breaking is that he sees both God and creatures as infinite. This new ground should not be underestimated. For it is rare for philosophers in the early modern period to hold, like Leibniz, that the distinction between an infinite God and finite creatures does not capture the complexity of this relation. To put it more bluntly, other than Leibniz, there is no major philosopher in the early modern period who argues that creatures are infinite.[2] The qualification of God as infinite beyond all determination may thus shed some light on his view of creatures as infinite *and* determinate at the same time.

---

[2] Spinoza's notion of infinite modes, of course, does come to mind. But it is not applicable to particular creatures, and the metaphysical framework is entirely different.

Antognazza articulates this point by arguing that, "Leibniz draws a fine line between being 'limited' and being 'finite.'" She elaborates on this as follows: "[s]trictly speaking, creatures for him are limited rather than finite since, through its confused perceptions, each individual substance involves the infinite" (20). Antognazza suggests that, since Leibniz clearly sees creatures as infinite, we should not describe them as finite but only as limited. Antognazza holds that "The difference between God and created substances ultimately reduces to one (but, to Leibniz's mind, crucial) difference: unlimited versus limited perfection" (19).[3] While I share this judgment, it seems to me that we can go further in substantiating and qualifying the curious status of created substance as limited *and* infinite. In line with Antognazza's observation that Leibniz's distinction between different kinds of infinity should be seen against a Neoplatonic mold, I will suggest that the distinction between God and created beings could also be clarified as a distinction between an absolute and a qualified degree of perfection and infinity, which also corresponds to degrees of being.[4]

One way to do this would be to look into Leibniz's notion of privation and negation as an imperfection, which would account for the limited nature of creatures—or, as I would put this, their limited degree of perfection.[5] Another way to substantiate this point would be to look at Leibniz's early distinction among three degrees of infinity (drawn in 1676) and consider to what extent we might apply this distinction to the relation between God (seen as the hypercategorematic infinite) and creatures, seen as infinite in a lesser degree. A third way would be to argue that Leibniz's notion of primitive force is related to the notion of the law of the series, such that both show the infinite and yet limited nature of individual substances. This will show that, whereas God is perfect and infinite without any qualification (i.e., absolutely), creatures are

---

[3] "Each soul knows the infinite—knows all—but confusedly" (PNG; AG 211).

[4] For Antognazza, the crucial feature of the Neoplatonic framework is that things are contained in God eminently and not formally. In addition, the Neoplatonic framework serves to cast the distinction between degrees of infinity as applying to different degrees of being. It is worth pointing out that, for Leibniz, all things are contained in God as possibilities. As Antognazza comments (in correspondence), "the very fact that creatures are contained in God eminently and not formally suggests that we have to do with possibility rather than actuality. In the classical, Christianized version of Neo-Platonism, the ideas of things as possibilities are embraced by God's intellect." For some elaboration on the Neoplatonic framework, see Antognazza's contribution to the 10th Leibniz Congress, "God, Creatures, and Neoplatonism in Leibniz," *Vorträge des X. Internationalen Leibniz-Kongresses*, ed. Wencho Li. (Hildesheim: George Olms, 2016), 3:351–64.

[5] As Antognazza notes, Leibniz collapses the traditional distinction between privation and negation.

infinite in some respects and limited in others. In what follows, each section develops one such way to argue for the status of created substance as infinite and limited.

8.2 PERFECTION AND REALITY

8.2.1 Perfection

This section presents some connections between infinity, perfection, and being. The next sections present the activity and passivity of living beings in regard to their relative imperfection (in comparison to God). I try to show that, according to Leibniz, the notions of infinity and perfection are related, such that they admit of similar degrees and hierarchical order.

First, let me call attention to the distinction between the notion of perfection (in the singular) and that of perfections (in the plural). Leibniz often uses the notion of perfections to describe the positive attributes of God and the properties of creatures. One of his early definitions of perfection appears in a note composed in Paris. It reads as follows:

> Perfection is an absolute affirmative attribute; and it always contains everything of its own kind, since there is nothing which limits it. ("Excerpts from Notes on Science and Metaphysics," March 18, 1676; A 6.3:392; LLC 53–55)

The context suggests that the notion of perfection Leibniz has in mind here is closely related to perfection as an attribute of God. Leibniz's definition of perfection is reminiscent of Spinoza's definition of an attribute: infinite in kind and distinct from the absolute infinity of God (which entails all attributes or perfections). Indeed, the connection with Spinoza is worth emphasizing, for the relation between perfection and reality is made explicit by Spinoza. In fact, this is an understatement, since for Spinoza this relation turns out to be an identity. As Spinoza states, "by reality and perfection I understand the same thing [*per realitatem, & perfectionem idem intelligo*]" (E IID6). Furthermore, there is a linear relation in Spinoza between degrees of perfection and degrees of reality, such that the more reality, the more perfection there is (E VP40).[6]

Leibniz's early definition of perfection suggests that there are various kinds of perfection—one corresponding to each quality or each kind of thing, where perfection is seen as a maximum (or the highest degree) of that quality or kind, such as wisdom and power. Again, the affinity with Spinoza's definition of the attributes and

---

[6] "*Quo unaquaeque res plus perfectionis habet, eo magis agit, et minus patitur, et contra, quo magis agit, eo perfectior est*" (EVP40).

the context of seeing perfections as the attributes of an infinite being seems fairly evident.

Some ten years later (1686), in the first paragraph of the "Discourse on Metaphysics," Leibniz articulates this view of perfection even more explicitly: "there are several entirely different ways of being perfect, . . . God possesses them all together, and . . . each of them belongs to him in the highest degree" (AG 35). As this passage shows, according to Leibniz, perfections (seen as attributes of God) admit of both variety and degree. In continuing the same paragraph, Leibniz articulates a very interesting test for finding out what a perfection is (although he actually ends up telling us what a perfection is not). He writes,

> [w]e must also know what a perfection is. A fairly sure test for being a perfection is that forms or natures that are not capable of a highest degree are not perfections, as for example, the nature of number or figure. For the greatest of all numbers (or even the number of all numbers), as well as the greatest of all figures, imply a contradiction, but the greatest knowledge and omnipotence do not involve any impossibility. (AG 35; A 6.4:1531)

As he also notes elsewhere,

> I believe it to be the nature of certain notions that they are incapable of perfection and completion, and also of having the greatest of their kind [*suo quoque genere summi*]. (A 6.3:551; LLC 179)

Leibniz's examples here are quite instructive. Knowledge and power are notions that can be regarded as perfections since they can have a highest degree, but number and figure cannot. In other words, maximal knowledge and maximal power do not involve a contradiction. This suggests that a notion that can have a highest degree (a maximum) can be regarded as a perfection. That Leibniz regards the notions of greatest knowledge and maximal power to be consistent, and that he contrasts them with the notions of the greatest number and the greatest figure, suggests that he does not consider them to be maximum in a quantitative sense. If wisdom and power were quantifiable in the same way as number and shape, they would also involve similar contradictions. This suggests that the sense in which Leibniz is using "perfection," and the sense in which it is considered infinite, in qualifying maximal knowledge and power, is not quantitative, either.

This should not be surprising, given that we are considering God's perfection and the perfection of his attributes. God is the most perfect Being. As we observed in chapter 2, he cannot be most perfect in a quantitative sense, for if he were, the notion would be contradictory.⁷ In this context, it seems rather clear that God's perfection is meant in an absolute and nonquantitative sense. In this absolute sense, perfection signifies a complete positive quality that cannot be measured by (or compared with) size, length, speed, or any other magnitude. Thus, the analogy between infinity and perfection becomes rather close to an identity in this context. Thus far, Leibniz's notion of perfection does not seem to be so different from Spinoza's. The absolute sense of perfection (in the singular) refers to the most perfect being, who is the subject of all perfections or positive attributes; in this sense, perfection and infinity are analogous: the most perfect being and the infinite being refer to one and the same thing—namely, to God—which is infinite and perfect in a metaphysical but not a mathematical (quantitative) sense. Infinity and perfection in kind apply in this way to the various perfections of God or his positive attributes, each of which refers to a certain quality (such as wisdom and power).

This sense of perfection, as an absolute and affirmative attribute of God, however, is not the only way in which Leibniz uses the notion of perfection. In some well-known texts, such as "On the Radical Origination of Things," the notion of perfection is used in a different way. Here, Leibniz defines perfection as a "degree of essence," so that the notion of degree is explicit in the very definition of perfection. It is remarkable that, in this context, perfection also serves Leibniz as the principle of existence—that is, the principle according to which God chooses to realize the most perfect world from all possible ones (GP VII:304). According to Leibniz, "the most perfect exists, since perfection is nothing but quantity of reality."⁸ Thus, God would choose the most perfect world—that is, the world whose degree of reality is the highest. Here are some textual variations that exemplify this notion of perfection: "Perfection is a degree of reality or essence [*perfectionem esse gradum seu quantitatem realitatis seu essentiae*]" (Leibniz to Eckhard, 1677; A 2.1:543);⁹ ". . . perfection being nothing but the degree or measure of positive reality [*la perfection n'étant autre chose que la grandeur de la réalité positive*] (GP VI:613); "Perfection is nothing but quantity of essence [*est enim perfectio nihil aliud quam essentiae quantitas*]" (GP VII:303).¹⁰

Implicit in these definitions is the idea already mentioned: the more reality (or essence or being), the more perfection. Leibniz uses this notion of perfection mainly

---

⁷ For details regarding this comparison, see chapter 2, this volume; and my "Leibniz on the Greatest Number and the Greatest Being," *Leibniz Review* 15 (2005): 49–66.

⁸ "Existe donc le plus parfait, puisque la perfection n'est autre chose que la quantité de réalité," in G. W. Leibniz, *Recherches générales sur l'analyse des notions et des vérités, 24 thèses métaphysiques et autres textes logiques et métaphysiques*, ed. and trans. J. B. Rauzy (Paris: Presses Universitaires de France, 1995), 469.

⁹ See L 178.

¹⁰ Perfection as harmonious coexistence with other things (A 6.4:1445; AG 19). This, Leibniz calls the "principle of perfection," or "the principle of the harmony of things" (A 6.3:472).

in the context of justifying God's choice of the best world (from all possible worlds). As he writes to Wolff,

> [t]he Perfection about which you ask is the degree of positive reality, or what comes to the same thing, the degree of affirmative intelligibility, so that something more perfect is something in which more things worthy of observation are found. (Leibniz to Wolff, Winter 1714–15, AG 230)

The degree of perfection of a world is the metric by which it is compared with other possible worlds; on this basis, the world with the highest degree of perfection can be chosen.[11] While the precise way in which Leibniz thinks about the relation between the perfection of individuals and the perfection of the world is very complicated, let us observe that, if possible worlds have different degrees of perfection, it is clear that this would also be expressed in their actual counterparts in the created world. In Leibniz's somewhat obscure terminology, the degree of perfection of each possible individual constitutes the degree of its claim for existence; the set of individuals whose actualization would realize maximal essence or perfection is then chosen for creation.[12] But the details of this complicated story need not concern us here.

8.2.2 God and Creatures

Leibniz writes that "the reality of creatures is not the same absolute reality that is in God, but a limited reality, for that is the essence of a creature" *(De abstracto et concreto*, 1688; A 6.4:990). In the *Monadologie*, Leibniz articulated this point in terms of perfection:

> 41. God is absolutely perfect—*perfection* being nothing but the magnitude of positive reality considered as such, setting aside the limits or bounds in the things which have it. And here, where there are no limits, that is, in God, perfection is absolutely infinite. (Theodicy §22)
>
> 42. It also follows that creatures derive their perfections from God's influence, but that they derive their imperfection from their own nature, which is incapable of being without limits. For it is in this that they are distinguished from God. (*Monadologie* 41, 42; AG 218)

---

[11] "And this reason [for existence] can only be found in fitness, or in the degree of perfection that these worlds contain, each possible world having the right to existence in proportion to the perfection it contains" (*Monadologie* 54. See also Theodicy §§74, 167, 350, 201, 130, 352, 345, 354).

[12] See, for example, "On the Ultimate Origination of Things" (GP VII:302–308; AG 149–55).

In this context, Leibniz is using perfection as a measure or as a degree of reality. God is the most perfect Being and "where there are no limits, that is, in God, perfection is absolutely infinite"; but creatures, having a particular and determinate combination of perfections, are particular (and thus partial) expressions of the divine essence; as such, they must have some limits (and hence some imperfections). For this reason creatures range lower on the scale of perfection. As Leibniz writes to De Volder,

> the supreme substance has poured forth his perfection as much as was permitted into the many substances that depend on him, which ought to be conceived of as individual concentrations of the universe and (some more than others) as imitations of the divinity. (January 1705; LDV 324–25)

Leibniz sees created substances as imitations of God—imitations with a lower degree of perfection. How are we to account for a lower degree of perfection in creatures, seen as imitations of the divinity? In other words, how can we account for imperfection in created substances? Created substances appear to share the same attributes as God, "[b]ut in God these attributes are absolutely infinite or perfect, while in the created monads or in entelechies ... they are imitations of it, in proportion to the perfection that they have" (*Monadologie* 48; AG 219). Note the explicit correspondence between infinity and perfection that Leibniz supposes here. As I indicated, I suggest that Leibniz supposes a similar correspondence with respect to degrees of perfection and infinity.

In the next paragraph (49), Leibniz relates perfection to activity and imperfection to passivity. Action is related to distinct perception, and passion (or being acted upon) to confused perception. Roughly stated, the main idea is that actions correspond to perfections and passions to imperfections. And these are related, respectively, to distinct and confused perceptions. Before considering this point, however, let me observe that, while an individual substance is (like God) an active being, an agent, and while it also "exerts infinite actions," Leibniz clarifies that it does not exert *all* actions. As he writes to Des Bosses:

> But it should not be thought on this account that, since it [substance] exerts infinite actions, it exerts every action whatsoever and every action equally, for each and every substance is of a determinate nature. (March 11, 1706; LDB 37)

Here, the determinate nature of creatures is explicitly invoked to account for the fact that, while each substance exerts infinite actions, it does not exert every action, since it is defined through a *determinate* and hence also a limited course of action. An individual's particular course of action can be seen through its particular sequence of perfections and imperfections. Since each individual consists of a unique sequence of perfections and imperfections, creatures differ from one another and all admit of some perfections and some imperfections—and yet, as we shall see (section 8.4), such a sequence is clearly infinite

8.2.3 Privation and Moral Imperfection

Let us briefly consider Leibniz's notion of privation as an illustration of imperfection. This in turn may help us to shed light on how these metaphysical imperfections can be the source of the moral imperfections Leibniz ascribes to intelligent creatures. The notion of perfections (in the plural) plays a major role here. According to Leibniz, every being is constituted and defined by virtue of a unique combination of perfections or positive attributes. Roughly speaking, we can think of this as a version of the complete concept of the individual. As I have argued at length (2007a, ch. 2), each such concept corresponds to a rule that generates a unique structure of predicates in God's understanding. The fact that such a structure is unique also reflects the fact that it is limited. Its uniqueness implies a particular combination of predicates that express (and indeed show) not only what it is but also what it is not. In this sense, every unique combination of perfections corresponding to a possible individual also shows or represents what the creature lacks, relative to the absolute perfection of the Creator (or perhaps more clearly, relative to all possible things, which God perceives in his understanding). This point can be made more explicit through the notion of logical space, seen as the space of all possible things. Seen against the space of all possibilities, the notion of a particular individual shows both the properties it has and the properties it lacks, which will translate into what it will do and what it will not do—that is, its actions and passions, in case it is created.

A curious illustration of this model of privation appears as Leibniz illustrates the creation by using a version of his binary notation, suggesting that the number 1 represents a perfection and 0 represents a privation (Grua 371).[13] In this context, Leibniz writes, "the privative is nothing but limits and there are limits everywhere in the creature just as there are points everywhere in a line" (Grua 126; cited in Antognazza 2009, 359). The complementary image of each unique concept (which corresponds to a complete concept of an individual) refers to the whole logical space

---

[13] "De compositione rerum ex perfectione et limitatione seu actu et privatione, deque origine rerum ex DEO et nihilo, ubi de origine numerorum ex 1 et 0 seu unitate et nullo" (Grua 371). See also Leibniz's letter to Fontenelle, February 26, 1701, in Antognazza (2009, 432).

of possible things, conceived in God's understanding. Ultimately, it refers to all the positive perfections, conceived in all possible manners by the divine mind—the famous region of ideas or realm of possible things. But each particular concept is necessarily limited, since it has a unique combination of perfections and thus as many privations, or perfections that it lacks, as points on a line. This privative nature of an individual may thus account for the "original limitation that the creature could not but receive with the first beginning of its being, and which is ultimately related to "the ideal reasons which restrict it" (Theodicy §§31, 141; See also *Monadologie* 50, 52).

If this model provides an illustration of what Leibniz might mean by privation, let us briefly consider how such a model, which fixes the positive and privative properties of creatures (their actions and passions), might serve as the source of moral imperfection. According to Leibniz, morality requires spontaneous action and rational judgment (which together with contingency constitute his three requirements for freedom). What could serve the transition from the notion of a fixed set of attributes or perfections into the realm of action is that we see the complete concept of an individual not merely as a combination of properties but also as a program *prescribing* (rather than causally determining) a particular course of action.[14]

Since any creature is defined through a certain program of action, it is still unclear what would distinguish rational beings from nonrational beings. The unique and therefore *limited* prescription of action of each creature may also account for the cognitive limitations imposed by the partiality and imperfection of such a being. The main difference is that a rational being has a capacity for reflection and judgment regarding his or her actions. In this reading, the individual's actions, which correspond to his or her predicates, need not be seen as the unmediated results of the individual's concept, as the results either of logical deduction or of mechanical execution; rather, rational individuals are seen as agents who act according to the reasons that incline them most. A rational agent thus would act according to what seems best to him or her.

The creature's limitation is expressed in this context through the way creatures employ the principle of the best. Unlike God, rational creatures do not act according to the absolutely best; rather, rational creatures act in accordance with what *seems* best to them. And it seems best to them from their limited point of view. This means that intelligent creatures act according to their unique but also limited point of view, as well as their limited cognitive capacities. It is perhaps in this sense that, for Leibniz, the metaphysical imperfections, seen as privations in the creatures' complete concept, may also account for a creature's moral imperfections.

[14] See Nachtomy (2007, ch. 6).

8.3 DEGREES OF PERFECTION AND THE THREE DEGREES OF INFINITY

In his *Meditations on First Philosophy*, Descartes assumes a very strong relation between infinity and perfection. Descartes contrasts the infinity and perfection of God with human finitude and imperfection. Indeed, the connection between infinity and perfection comes close to an identity insofar as Descartes uses these terms interchangeably. Leibniz's analogy between perfection and infinity (imperfection and limitation), however, is not restricted to one sense of infinity—namely, to the absolute sense of infinity, which pertains to the *Ens Perfectissimum* alone. Unlike Descartes and most other thinkers in the early modern period, Leibniz is not merely distinguishing between the absolute reality and perfection of God and the limited and imperfect nature of creatures. This distinction certainly holds for Leibniz. But, as we have seen, for Leibniz, creatures are imperfect with respect to the absolute perfection of God.

In notes from 1676, Leibniz distinguishes explicitly among different degrees of infinity. So, let us consider the question Antognazza also takes up (2015, 16–18), viz., how the distinction between degrees of infinity (drawn in 1676) is related to the conception of infinity Leibniz outlines in 1706.

In his summary of Tschirnhaus's account of Spinoza's philosophy, Leibniz writes:

> He [Spinoza] thinks that there are infinitely many other affirmative attributes other than thought and extension, but that there is thought in all of them, as there is here in extension; but that we cannot conceive what they are like, each one being infinite in its own kind, as, here, is space. (A 6.3:385; LLC 43)

Leibniz clearly refers here to Spinoza's distinction between the absolute infinity of God and the infinity in kind of its attributes. This distinction is one of the first definitions of the *Ethics*, and it is likely that Leibniz heard about it from Tschirnhaus (since he had no access to the *Ethics* at this point). Immediately following this remark, Leibniz makes this remarkable note:

> I usually say that there are three degrees of infinity. The lowest is, for the sake of example, like that of the asymptote of the hyperbola; and this I usually call the mere infinite (*tantum infinitum*). It is greater than any assignable, as can also be said of the other degrees. The second is that which is greatest in its own kind (*maximum in suo scilicet genere*), as for example the greatest of all extended things is the whole of space, the greatest of all successives is eternity. The third degree of infinity, and this is the highest degree, is *everything*

(*omnia*), and this kind of infinite is in God, since he is all one; for in him are contained the requisites of existing of all others. (February 1676; A 6.3:386; LLC 43)[15]

As Antognazza acknowledges, the highest degree of infinity fits quite neatly with Leibniz's view of God as the hypercategorematic infinite (2015, 17). The lowest degree of infinity clearly applies to numbers and other concepts produced by addition (or composition) of parts, understood syncategorematically. But what is Leibniz's intended use of the second degree of infinity? As Antognazza judiciously observes, "in the context of notes commenting on Spinoza such as these, Leibniz is simply thinking (as Spinoza) of the infinity of the divine attributes" (18). Indeed, in his annotations to Spinoza's letter, Leibniz makes these same distinctions even more explicit (though in reversed order). He writes:

> I set in order of degree: *Omnia*; *Maximum*; *Infinitum*. Whatever contains *everything* is maximum in entity; just as a space unbounded in every direction is maximum in extension. Likewise, that which contains everything is most infinite, as I am accustomed to call it, or the absolutely infinite. The *maximum* is *everything* of its kind (*omnia suis generis*) i.e. that to which nothing can be added, for instance, a line unbounded on both sides, which is also obviously infinite for it contains every length. Finally those things are *infinite in the lowest degree* whose magnitude is greater than we can expound by an assignable ratio to sensible things, even though there exists something greater than these things. (A 6.3:282; LLC 115)

If an analogy between degrees of perfection and degrees of infinity can be developed along the lines noted here, I would suggest that creatures may be seen under the intermediate degree; for they realize a maximum in kind, the particular and determinate way of being that defines their unique essence and nature (i.e., the unique sequence of predicates expressed by their complete concept). As we have seen in the previous section, unlike God, creatures have some perfections, but also many

---

[15] "Ego Soleo dicere: tres infiniti gradus, infimum v.g. ut exempli causa asymptoti hyperbolae; et hoc ego soleo tantum vocare infinitum. Id est majus quolibet assignabili; quod et de caeteris omnibus dici potest; alterum est maximum in suo scilicet genere, ut maximum omnium extensorum est tosum spatium, maximum omnium successivorum est aeternitas. Tertius infiniti, isque summus gradus est ipsum, omnia, quale infinitum est in Deo, is enim est unus omnia; in eo enim caeterorum omnium ad existendum requisita continentur. Haec obiter annoto."

limitations and imperfections.[16] To be sure, they have all the perfections they can have—and are maximally consistent in this sense—but this of course falls short of having all perfections in a nonqualified way (which is reserved for God alone). While, in the context of these 1676 notes, Leibniz may well have Spinoza's definition of an attribute in mind, it remains beyond doubt, as Antognazza observes, that, "for Leibniz, creatures are (in more than one sense) infinite" (2015, 23).

Indeed, as Antognazza emphasizes, Leibniz articulates by the distinction between God and creatures in a rather unusual way. In a passage from a letter to Bayle (ca. 1702) that Antognazza cites, Leibniz writes:

> I don't know whether it is possible to explain the constitution of the soul better than by saying (1) that it is a simple substance, that is, what I call a true unity; ... (4) that the soul is an imitation of God as much as is possible to creatures; that it is, like Him, simple and yet also infinite, and envelops all through confused perceptions, but is limited as regards distinct perceptions. Whereas everything is distinct for the sovereign substance, from which everything emanates, and which is the cause of existence and of order, and in a word, is the ultimate reason of things. God contains the universe eminently, and the soul or the unity contains it virtually, being a central mirror, but active and vital, so to speak. One can even say that each soul is a world apart, but all these worlds harmonize and are representative of the same phenomena reported in a different way. (GP III:72)[17]

Leibniz is very explicit here in stating that creatures are simple, infinite, and limited. Is there any connection between the intermediate degree of infinity—a maximum in its kind—and the infinite and limited nature of creatures (which allows for a degree of perfection and infinity)? I suggest that the notion of maximum in kind could be used in this context, so that a creature is seen as both infinite and limited, perfect and imperfect, such that it has both perfections and imperfections. As we have seen, Leibniz's view is articulated in terms of the notion of action, viewed as a realization of perfection and passion as a lack of perfection and thus as an imperfection. Roughly

---

[16] "The imperfection that remains in them [creatures] comes from the essential and original limitation of created things" (PNG; AG 210). "The imperfections, on the other hand, and the defects in operations spring from the original limitation that the creature could not but receive with the first beginning of its being, through the ideal reasons which restrict it. For God could not give the creature all without making of it a God; *therefore there must needs be different degrees in the perfection of things and limitations also of every kind*" (Theodicy §§31, 141–42, my emphasis). For an illuminating discussion of the limitation of creatures, see Antognazza (2015).

[17] For an analysis of this passage, see Antognazza (2015, 19–20).

stated, what a creature does is considered a perfection; when a creature suffers or receives an action (when it functions as a patient), it is a passion and is considered an imperfection.[18] And it goes without saying that for anything the agent does, there are so many things it doesn't do.

Now, recall that the very definition of a creature is to have a particular and determinate course of action, illustrated by its complete concept such that it includes all its proper predicates. When the notion of a maximum in kind is seen through Leibniz's notion of a complete concept of an individual, which prescribes a unique course of action expressed by a unique sequence of predicates, then in comparison with the whole of logical space, such an individual notion permits a view not only of what the individual is but also of what it is not. This picture turns out to imply both what the individual will do (its actions, corresponding to its perfections) and what it will not do (its passions, corresponding to imperfections). In this sense, the individuality and uniqueness of each thing turn out to reflect both what it is and what it is not—what it will do (as an agent) and what it will suffer as a recipient of action (patient).[19] In having a particular sequence of perfections and imperfections, the individual's nature thus admits of a lower degree of perfection and infinity. It is maximal in its kind, but it is also a particular (and individual) determination and, as such, it is also limited.

8.4 INDIVIDUAL SUBSTANCE, FORCE, AND THE LAW OF THE SERIES

It is well known that, for Leibniz, every individual substance acts, and everything that acts is an individual substance. This is how Leibniz reads the dictum *Actiones sunt suppositorum*—actions pertain to subjects, or as Ariew and Garber render this, "actions are of individual subsistent substances" (AG 40n73). Actions pertain to individual substances, but as Leibniz clarifies, this also holds reciprocally, so that anything that acts is an individual substance—a subject of action or an agent. Note that this view supposes a fixed source of actions, so that all actions are indeed *of* (or belong to) the same individual or the same agent.[20] Indeed, as Leibniz writes,

> active force involves an effort (*conatus*) or striving (*tendentia*) toward action (*actio*) so that, unless something else impedes it, action results. And, properly speaking, entelechy, which is insufficiently understood by the schools,

---

[18] "Perfection is pure act or pure positivity. What we commonly say of act and potency is more correctly said of the positive and privative, or of the absolute and limited" (ca. 1695–1697; Grua 371, cited in Antognazza 2015, 30n68).

[19] In this sense, a created substance may also be seen as a living mirror of the whole universe.

[20] As Russell (1992, 40) pointed out, this implies that activity is the very essence of substance.

consists in this.... Primitive force, which Aristotle calls first entelechy and one commonly calls the form of a substance, is another natural principle which, together with matter or passive force, completes a corporeal substance. ("Against the Cartesians"; GP IV:395; AG 252)

The point I emphasize here is that the source of action or the primitive force of each individual substance, if adequately seen in the metaphysical context of discussing forces, is not a finite quantity. This may seem initially surprising. For Leibniz assigns to created beings a force that always remains the same; it is the very feature of nature that remains invariant (both in each substance and in the whole of nature) in the midst of constant variations, such as motion, modifications, and other change in properties. As is well known, Leibniz's main critique of Descartes's conservation law is that it is not only empirically inadequate but also misses the very aspect of nature that remains fixed and invariant: it is not the quantity of motion (size multiplied by speed) that remains the same, Leibniz argues, but, rather, its source—that is, motive force, measured by mv^2. Operating in the background here is Leibniz's critique of Descartes's attempt to describe the natural world in purely geometrical terms. Replacing extension as the essence of body with force is part of Leibniz's attempt to resist the reduction of nature to mere extension, and thus to a pure mechanism that can be described in purely quantitative terms. In this context, however, since primitive force always remains the same, the force Leibniz ascribes to each individual would seem to be fixed and finite.

A closer look, however, reveals that, precisely because it remains fixed, primitive force may be best regarded as infinite (although the sense in which it can be called infinite would have to be qualified as a metaphysical notion). In a letter to De Volder (January 1704), Leibniz writes:

> You say that motion, i.e., that which results from size and speed, is derivative force. I do not take motion to be derivative force, but I do think that motion (namely change) follows from it. Moreover, derivative force is *the present state itself* in so far as it tends toward a following state, i.e., pre-involves a following state, *just as everything in the present is pregnant with the future*. But the persisting thing itself, insofar as it involves all cases, has primitive force, so that primitive force is like the law of a series, and derivative force is like a determination that designates some term in the series. (LDV 287; GP II:262, my emphasis)

According to Leibniz, primitive and derivative forces are related as invariable and variable, or as a substance and its modification.[21] As Robert Adams notes,

---

[21] "And indeed, derivative forces are nothing but modifications and echoes of primitive forces" (LDV 263).

"primitive and derivative forces are related as persisting subject and [its] present state" (1994, 379).[22] Most pertinent to my purposes here, primitive force is likened to the law of the series, and derivative force is likened to a determination that designates some term in the series. Roughly speaking, primitive force corresponds to the principle of change (which persists), and derivative force corresponds to the actual state of the substance (designated by the current term in the series); primitive force refers to the law that generates change, and derivative force to the present state. Primitive force is constant and continuous, while the various states are temporal and discrete. Likewise, the law is one and its states are varying and proceed to infinity.[23]

The way Leibniz conjoins primitive force and the law of the series merits some development. I believe that Leibniz integrates two aspects here: the law provides a course of action, and primitive force provides power of action. The law provides the information as to what to do, and the force provides the power that enables its execution. These two aspects are necessary for Leibniz's view of individual substances. In itself, the law can be seen as pointing to a possible individual (conceived in God's mind), while the conjunction of law *and* primitive force can be seen as pointing to an actual (created) individual.

The point I highlight here, however, is that the notion of force Leibniz employs to define an individual substance has a qualitative aspect that cannot be reduced to size and speed (as the motions that result from it may). In a letter to Bayle, Leibniz writes,

> I would like to add a remark of consequence for metaphysics. I have shown that force ought not to be estimated by the product of speed and size, but by the future effect. However, it seems that force or power is something real at present, while the future effect is not. From this it follows *that we must admit in bodies something different from size and speed, at least unless one wants to refuse bodies all power of acting.* (Leibniz to Bayle, January 9, 1687; GP III:48, cited in Garber 1995, 287)

Power of acting or primitive force seems to have a qualitative side, since the ability to do work in the future is real at present, but at the same time it cannot be fully described in quantitative terms.[24] Since this may sound a bit mysterious, let me highlight this point from another angle. Let us reflect on the fact that primitive force,

---

[22] For further illuminating discussion of the relation between primitive and derivative forces, see Adams (1994, 379–80); Garber (1995, 292–93).
[23] See Phemister (2005, 195).
[24] For an elaboration of this point, see Garber (1995, 287).

despite functioning as the continuous cause of motion and change, does not ever increase or decrease, as it constitutes the cause of further successive changes. In fact, this is precisely the point of arguing that primitive force is preserved through changes of state. On the face of it, any physical force that is the cause of change and is constantly active ought to be diminished. But this would be true if such a force is considered as a quantity. But, since primitive force does not diminish, this suggests that, for Leibniz, primitive force is not conceived as a finite quantity—or, better, it is not conceived as a quantity, since any quantity is finite.

If so, primitive force for Leibniz should not be understood through the model of a reservoir that loses water as it continuously flows out of it; nor should it be seen as a fountain that is being filled while it flows, nor as a source of energy that is being recharged while it is used. That is to say, the preservation of primitive force is not maintained by an equality of what is lost and what is gained, or by what it gives and what it receives. Rather, Leibniz's point is that, while primitive force constitutes a continuous cause of motion (and, more broadly, a source of action), it does not lose anything as it serves as the source of future actions. It need not be recharged because it never loses anything.

But how can this be? Indeed, we need to recall that primitive force (like the principle of life) derives from a divine source. As the divine power itself remains the same and never diminishes, so is the case for primitive individual forces created by God: these are created individual forces that serve as divine agents on earth, so to speak (and in the dual sense of "agent"). Compare this with the Cartesian picture of the source of motion in the world. According to Descartes, the quantity of motion is conserved because motion stems directly from a divine source—a source that must be seen as immutable. In Leibniz, the source of motion in the universe also remains the same, but this derives not directly from God but from God's creating many individual sources of action, each of which is ever active and never diminishing.

As Leibniz states in the following passage, active force ought to be attributed to the force God placed in things:

> many things force us to place active force in bodies, especially experience itself which shows that motions are in matter. Though its origin ought to be attributed to God, the general cause of things, however, directly and in particular cases, they ought to be attributed to the force God placed in things. For to say that, in creation, God gave bodies a law for acting means nothing, unless, at the same time, he gave them something by means of which it could happen that the law is followed, otherwise, he himself would always have to look after carrying out the law in an extraordinary way. But indeed, this law

is efficacious, and he did render bodies efficacious, that is, he gave them an inherent force. (1702; AG 253–54; GP IV:396–97)

While the ground for the conservation of primitive force is its divine source, created beings are efficacious owing to their inherent force; they are agents endowed with a divine-like source of action—primitive active force—that does not diminish as long as they act. Once created, substances always act, but their source of action does not decrease. In this sense, primitive force does not seem to be a quantity, for otherwise it would be finite and would diminish. But since it does not change—neither decreases nor increases—it may be better seen as infinite in a qualitative and metaphysical sense. At the same time, the power of created things is obviously not the same as the absolute power (omnipotence) of God; rather, while infinite, it must be regarded as infinite in a lower degree. Here is additional textual evidence for this view: "Each substance has something of the infinite insofar as it involves its cause, God: namely some vestige of omniscience and omnipotence" ("A Specimen of Discoveries," 1686?; A 6.4:1618; LLC 309). This vestige of omnipotence is what I emphasize here. Creatures are obviously not omnipotent, but they do have some infinite degree or measure of power. A creature's power is, of course, limited, since it is individual and thus also particular. As Leibniz writes,

> every created thing contains both the limited and the unlimited: the limited in respect of distinct cognition, and of irresistible power, and the unlimited in respect of confused cognition and of diffused action. For every soul, or rather every corporeal substance, is confusedly omniscient and diffusely omnipotent. For nothing happens in the whole world which it does not perceive, and it has no endeavor that does not extend to infinity. ("Wonders Concerning the Nature of Corporeal Substance," 1683; A 6.4:1465–66; LLC 265)[25]

While these passages require further clarification, it is clear, I think, that primitive force, which together with a law of the series constitutes a created substance, does not change and may therefore be regarded as infinite in the sense of producing a constant

---

[25] See also "The power of every body is infinite. Now I call a body one if every action of its parts is an action of that one body and if the parts of this body are infinite. And so the infinite force, which is contained by an equal or even greater striving" (A 6.4:1401; LLC 249). "*Individuals (singularia)* involve infinity; in forming *universals* the soul only abstracts certain circumstances by concealing innumerable others. Ans so it is only in an individual that there is a notion so complete that it also includes all of its changes" (Leibniz to De Volder; N64, 1705; LDV 325).

and undiminishing source of action.[26] It is a universal power of action made individual through its connection with a particular law of a series: "the persisting thing itself, insofar as it involves all cases, has primitive force, so that primitive force is like the law of a series" (LDV 287; GP II:262).[27] In this sense, each individual may aptly be seen as an imitation of God in that it involves a lower degree of infinity. This degree of infinity may well be seen along the lines of the intermediate degree of infinity Leibniz calls a maximum in kind in his notes from 1676. And in line with the Neoplatonic mold that Antognazza stresses, these degrees of infinity correspond to degrees of perfection and being.

8.5 CONCLUSION

Assuming Leibniz's characterization of the most perfect being as infinite in a hypercategorematic sense—that is, a being which is beyond any determination—I suggested that creatures may be distinguished from God in constituting determinate and particular expressions of the divine essence. Given that, for Leibniz, both God and creatures are infinite, this leads to characterizing creatures as both infinite and limited. I offered three lines of argument to substantiate this point: (1) seeing creatures as entailing a particular sequence of perfections and imperfections; (2) seeing creatures under the rubric of an intermediate degree of infinity and perfection that, in 1676, Leibniz calls maximum in kind; and (3) observing that primitive force, a defining feature of created substance, may be seen as infinite in a metaphysical sense. This leads to viewing Leibniz's use of infinity within a Neoplatonic framework of descending degrees of being: from the hypercategorematic infinite, which is identified with the most perfect being; to the lowest degree of *entia rationis* (or beings of reason). In this framework, created substances occupy the curious intermediate position, as they are both infinite and determinate. In stark contrast to the hypercategorematic infinity of God, which is above any determination and limitation, Leibnizian creatures are defined by a particular sequence of predicates; in this way, they are determined and limited and yet infinite in kind.

---

[26] In referring to GP IV:395; AG 252, Garber writes that "the primitive forces, active and passive, come together to make up the corporal substance, the genuine unity that, Leibniz claims, underlies the extended bodies of physics" (Garber 1995, 291).
[27] "The situation is as it is with the laws of series or the natures of curves, where the entire progression is fully contained in the very beginning. The whole of nature must be like this: otherwise it would be absurd and unworthy of wisdom" (Leibniz to De Volder, November 19, 1703; LDV 279).

9

MONADS AT THE BOTTOM, MONADS AT THE TOP, MONADS ALL OVER

9.1 INTRODUCTION

In his extensive essay "L'invention métaphysique," published as an introduction to his edition *G. W. Leibniz: Discours de métaphysique suivi de Monadologie et autres textes*,[1] Michel Fichant provides a reconstruction of Leibniz's transition from the metaphysics of the "Discourse on Metaphysics" (1686) to that of the *Monadologie* (1714). In this essay, Fichant makes the following remark:

> From a monadological point of view, the complete notion of an individual... is replaced by an organic complex, or an infinite multitude of monads, which is subordinate to a dominant monad—a monad that constitutes [*qui en fait*] the entelechy or the primitive force [of the whole complex]. (2004, 130)

I would like to highlight two points here. First, according to Fichant, a dominant monad constitutes the entelechy or the primitive force of the whole complex—that is, it constitutes the source of activity of an organic complex. Second, as Fichant observes, it is the *complete notion of the individual* that is being replaced by an organic complex, and it is this complex that is subordinate to a (single) dominant monad. If Fichant is right (and I think he is), this suggests that Leibniz's notion of the whole individual substance that dominated the years of the "Discourse" takes a new form within the monadological framework. And this suggests that the notion of an individual substance still figures—albeit transformed—within the framework of the *Monadologie*, a framework that, if seen as consisting of monads at the bottom level alone, has no place for the notion of an individual substance such as a human being or an animal. One indication of this apparent lack of individuality is that monads have become universal: in replacing the picture of the "Discourse," in which individuals bear proper

---

[1] Fichant (2004).

names (such as Alexander and Julius Caesar), monads are nameless, and the term (monad) seems to range over all sorts of true units.

The picture of the *Monadologie* is indeed complex and bewildering; for one thing, it involves a hierarchy of many (in fact, infinitely many) monads that are somehow supposed to make up a single entity, even if the monads are to be regarded as causally independent of one another and even if aggregates are clearly distinguished from true substances. Whereas a monad is simple and without parts, Leibniz is committed to the claim that "this diversity (which is essential for a monad) must involve a multitude in the unity or in the simple" (*Monadologie* 13; AG 214). Leibniz's point is that a simple monad involves a "plurality of affections and relations, although it has no parts" (*Monadologie* 13). This needs to be emphasized because it goes against the most pervasive reading of the *Monadologie* and related texts—a reading which is indeed suggested by the opening paragraphs of the *Monadologie* and the "Principles of Nature and Grace" (PNG). The common reading I am referring to is the following: monads are to be found only at the bottom level of reality, and they enter into the composition of complexes (or composites) as elementary constituents or building blocks; the resulting complexes in turn are seen as aggregates of simple substances. As Leibniz clarifies, "the composite is nothing more than a collection or *aggregate*, of simples" (*Monadologie* 2).

This reading is clearly articulated in Daniel Garber's *Leibniz—Body, Substance, Monad*. As he puts it, "[w]hatever else there might be—be they bodies, corporeal substances, composite substances or whatever—they must ultimately be grounded in some way in simple substances or monads."[2] For monads are "the true atoms of nature and, in brief, the elements of things" (*Monadologie* 3).[3] Garber argues that, in contrast to the previous corporeal substance metaphysics, the monadological picture presupposes partless and nonextended basic units—monads—as the grounds for whatever else is real in the world, and whatever else there might be must be *grounded* in the monads. I take this bottom-up grounding picture to imply that whatever else there is, is composed of monads.[4]

---

[2] Garber (2009, 317). For a more recent statement of this view, see Jorati (2017, 9).

[3] For a recent and explicit analogy between monads and atoms, see the entry for "Monad" in the *Routledge Encyclopedia of Philosophy* by Jen Nguyen and Jeffrey McDonough. See *Routledge Encyclopedia of Philosophy* (Taylor and Francis, 2017). https://www.rep.routledge.com/articles/thematic/monad/v-1.

[4] In a more recent essay ("Monads on My Mind," in *Leibniz's Metaphysics and Adoption of Substantial Forms*, ed. Adrian Nita [Dordrecht: Springer, 2015]: 161–76), Garber makes this reading more precise by looking closely at the emergence of monads between 1695 and 1698.

In what follows I argue that the scope of Leibniz's use of monads is broader, such that they are to be found not only at the metaphysical bottom level but also at the top. That is, a grounding relation goes not just from the bottom to the top but also from the top to the bottom.[5] And, as I shall also argue, these are not only different directions but also different senses of grounding. The grounding of the dominant monad from top to bottom is active; in conferring unity and functional organization, the dominated monad gives rise to complete substances or organic beings. The grounding from the bottom up serves to ensure that even nonliving beings are, ultimately, made up of some living, indivisible units—or some kind of true beings that guarantee the reality of bodies.

9.2 INCOMPATIBLE STRANDS IN THE MONADOLOGICAL TEXTS?

To get a better grip on the origin of the widely accepted reading (that monads are only at the bottom), let us attend to the opening paragraph of the "Principles of Nature and Grace":

> A substance is a being capable of action. It is simple or composite. A *simple substance* is that which has no parts. A *composite substance* is a collection of simple substances or *monads*. *Monas* is a Greek word signifying unity, or what is one. Composites or bodies are multitudes; and simple substances – lives, souls, and minds—are unities. There must be simple substances everywhere, because without simples, there would be no composites. As a result, all of nature is full of life. (PNG §1; AG 207)

It is evident that the picture of simple substances occupying the bottom level indeed looms large here. However, the picture of simple substances at the bottom level alone conflicts with another feature, which is also evident in the same texts, as well as in other texts from this period. This is particularly manifest in Leibniz's frequent examples of substances and true unities as the Ego, the self, the animal, and, more explicitly, a complete animal. Before looking at these examples, let us observe that what Leibniz calls composite substance in the opening paragraphs of the *Monadologie* and the PNG is very closely related to (if not identical with) what he usually regards as

---

[5] The common presumption is that "atom" refers to the elements from which complex substances are composed. But this is not the only way of thinking of a Leibnizian atom. Unlike Garber, Pauline Phemister holds that the "true atoms of nature" point to corporeal substances in the monadological phase as well; see Phemister (2005, chs. 2 and 3). See also P. Phemister, *Leibniz and the Environment* (London: Routledge, 2016), 38–39.

aggregates—that is, collections of substances that are *not* themselves considered to be true unities but, rather, are collections of such unities. Hence, Leibniz's characterization of a *composite substance* as "a collection of simple substances" is in tension with his usual definition and usage of the term "substance," in clear distinction from an aggregate.

For example, in a letter to De Volder, Leibniz says explicitly that, in contrast to an aggregate, a monad, *with its body*, makes up a *complete* substance:

> What I take to be the indivisible or complete monad is the substance endowed with primitive power, active and passive, like the "I" or something similar. (Leibniz to De Volder, June 20, 1703; GP II:251; AG 176)[6]

Note that the second strand is evident in the PNG and *Monadologie* as well. For example, Leibniz writes the following:

> Each monad, together with a particular body, makes up a living substance. (PNG §4; AG 208; GP VI:599)

> The body belonging to a monad (which is the entelechy or soul of that body) together with an entelechy constitutes what may be called a *living being*, and together with a soul constitutes what is called an *animal*. (*Monadologie* 63; AG 221)[7]

This point is also expressed with remarkable clarity in a letter to Johann Bernoulli in 1698:

> (4) What I call a complete monad or individual substance [*substantia singularis*] is not so much the soul, as it is the animal itself, or something analogous to it, endowed with a soul or form and an organic body.

> (5) You ask how far one must proceed in order to have something that is a substance, and not [a collection of] substances. I respond that such things present themselves immediately and even without subdivision, and that

---

[6] Commenting on this letter, Garber writes that "it now seems that Leibniz is conceiving of the unities that ground his world on the model of souls rather than animals," Since "monads or simple substances are non-extended," these monads, rather than the corporeal substances of the middle period, now constitute the ultimate grounds of reality (Garber 2009, 312).

[7] See also the letter to Bierling, 1711: "I call corporeal substance that which consists in a simple substance or monad (that is, a soul or soul-analogue) and a united organic body" (GP VII:501, cited in LDV liii).

every animal is such a thing. For none of us is composed of the parts of our bodies. (September 20/30, 1698; AG 168; GM 551–52)[8]

This passage is very telling. Note that Leibniz speaks of "a complete monad or individual substance" as if these were interchangeable terms and identifies it with the animal itself—a soul or form *and* its organic body.[9] The next paragraph is even more instructive. How far does one need to proceed in order to get to the fundamental level of substances (rather than collections of them)? Leibniz's response is revealing: not at all, for substances are already found at the top—that is, they "present themselves immediately . . . and every animal is such a thing." And, once again, Leibniz is using his favorite example of the self or a human being for a complete substance, arguing that *we* are not composed of the parts of our bodies. That we are not composed of the parts of our body implies that, *qua* complete substances, we are without parts. Or, to put this differently, the model of parts and whole does not apply to a complete substance such as a human being or an animal; it applies only to its body, regarded as such. Thus, the picture of composition would apply only to aggregates or wholes that admit of parts. True units or monads are, however, without parts and hence are not such wholes. Rather, the model of parts and wholes does not apply to monads or true beings. True beings have no parts and, for that reason, are indivisible.

9.3 RECONCILING MONADS AT THE BOTTOM WITH MONADS AT THE TOP

In light of these passages, I would like to raise the following question: Can we reconcile these two evident (but seemingly opposing) strands in Leibniz's later texts? The one strand identifies substances as the fundamental building blocks that enter into the composition of complexes, and the other strand identifies substances with complete substances such as animals or humans.

The unity of body and monad, seen as the soul or entelechy of that body, is further emphasized in the third paragraph of the PNG, where Leibniz writes, "a body [belonging to a central monad] is organic when it forms a kind of automaton or natural machine, which is not only a machine as a whole, but also in its smallest distinguishable parts" (AG 207; GP VI:599). Note that in these passages, the notions of living being and organic body are playing a crucial role. The identification of a living body with a natural machine—a strand that originates with the "New System

---

[8] For a similar conception of the monad from 1698, see *On Nature Itself* (GP IV:511; AG 162). For a late reference to a similar point, see the letter to Remond of November 4, 1715 (GP III: 657 and 658).
[9] See Phemister (2005, 72) for discussion.

of Nature"—is clearly articulated in the *Monadologie* as well: "Thus each organic body of a living being [*vivant*] is a kind of divine machine or a natural automaton, which infinitely surpasses artificial automata. For a machine constructed by man's art is not a machine in each of its parts" (*Monadologie* 64; AG 221, translation slightly modified).[10]

As we saw in chapter 6, Leibniz's notion of a natural machine (in contrast to that of an artificial machine) serves to distinguish living beings from nonliving beings. In this context, an artificial machine is seen as a collection (or an aggregate), and a natural machine is seen as an organized single unity. A natural machine involves a hierarchical structure of machines that remain machines to the least of their parts, governed by the domination relation. This picture implies that, while natural machines are nested within each other *ad infinitum*, complete living beings activated by their own entelechies are found at every level. Thus, for Leibniz, every aspect of nature—all the way down to infinity—is also full of life (*Monadologie* 66–70; PNG §1). At the same time, it is also clear that animals (large and small) remain one of Leibniz's frequent and paradigmatic examples of complete substances. In other words, the Leibnizian world seems to be populated with animals (or, more generally, living beings) both at the bottom *and* at the top, both at the invisible microscopic level *and* at the visible level of what we commonly designate as animals in everyday life.

Indeed, it seems that a similar picture is expressed in the framework of the *Monadologie*. Thus, we might further speculate that (together with the intensive use of force[11]) the notion of natural machine, developed first at 1695, plays an important role in the transition from the "Discourse"'s metaphysical picture with its complete individuals to the universalized notion of a monad we find in the *Monadologie*. This point is further supported by the passage that was partially cited earlier:

> Each monad, together with a particular body, makes up a living substance. Thus, there is not only life everywhere, joined to limbs or organs, but there are also infinite degrees of life in the monads, some dominating more or less over others. (PNG §4; AG 208)

---

[10] Note that Leibniz identifies here the organic body with the natural machine rather than the whole animal. There are enough texts to justify seeing the whole animal or substance as a natural machine. But this is a controversial point. For example, François Duchesneau, Justin Smith, and Richard Arthur tend to prefer a less radical reading in which a natural machine corresponds to the organic body of the whole animal rather than to the animal itself. See Duchesneau (1998); Smith (2011); Arthur (2014).

[11] See Phemister (2016, 37).

Let us connect this point about infinite degrees of life in the monads with the hierarchy of perfection that permeates the created world. As Robert Adams observes, "[t]he mature Leibniz seems to have maintained that it is precisely in degrees of perfection, in various respects, that finite substances differ from each other and from God."[12] If, for Leibniz, "Domination and subordination for the monads consists only in degrees of perfection" (Leibniz to Des Bosses, June 16, 1712; GP II:451; LDB 257), this hierarchy of perfection may well correspond to the degrees of life and activity in the monads.[13] What is crucial here is the correlation between the monads' degrees of life and the hierarchical order implied by the domination relation in which the monads are ordered. The domination relation corresponds to degrees of activity, so that the more dominant a monad, the more active it is. And this degree of activity corresponds to the monad's degree of perfection.[14]

As I have argued in chapter 6, this order is an essential feature of Leibniz's notion of a natural machine, in the context of which a higher degree of activity means a higher degree of functional organization.[15] This notion of hierarchy in the organic (or living) aspect of nature is clearly expressed in section 70 of the *Monadologie*:

> Thus we see that each living body has a dominant entelechy, which in the animal is the soul; but the limbs of this living body are full of other living beings, plants, animals, each of which also has its entelechy, or its dominant soul. (*Monadologie* 70; AG 222)[16]

In section 63, Leibniz says that the monad of a body "is the entelechy or soul of that body." For this reason, it seems perfectly legitimate to think that in section 70 he has dominant monad in mind when he talks about a dominant entelechy and a dominant soul. Still, there is a significant difference between saying that "the limbs of this living body are full of other living beings" and saying that a living being is organized in a tight net of functional relations between its constituents. Unfortunately, Leibniz often conflates these two claims. But generally speaking, the claim that there are creatures within a living being is not sufficient to distinguish between a

---

[12] Adams (1994, 122).

[13] There is some doubt here whether Leibniz meant perfection or perception in the cited letter. As Look and Rutherford clarify, "in the draft Leibniz wrote 'degree of perfection [*gradibus perfectionum*].' The sent version 'degrees of perception' may be a copyist error" (LDB 441).

[14] "[I]n as much as we follow this perfection of our nature [reason], we are said to act (*agir*), . . . and in so far as we are imperfect, we are said to suffer (*patir*)" (Grua 481).

[15] See also Brandon Look, "On Monadic Domination in Leibniz's Metaphysics," *British Journal for the History of Philosophy* 10, no. 3 (2002): 379–99.

[16] See also A II.3:341.

true substance and an aggregate consisting of a collection of substances. For, according to Leibniz, anything whatsoever contains living beings, be it a substance or an aggregate.[17]

Another facet of responding to the question of how to reconcile the presence of substances both at the top and at the bottom comes up in another insightful observation by Fichant: "the extension of the concept [monad] allows for the possibility of a qualitative variation, since there are degrees of unity, where the organic composition of bodies [cor]responds to the hierarchy of monads" (2004, 136). While there is a dominant entelechy (or a principle of action) at the top of each animal, there is also a whole hierarchy of dominated entelechies (one governed by the other, so to speak) *ad infinitum*. And this nested structure of dominating and dominated entelechies is, as already noted, an essential aspect of Leibniz's view of living beings.[18] In fact, it is this very complex system of functionally organized machines from the top down that accounts for the distinction between living and nonliving things. A living being, according to Leibniz, does not merely contain other living beings (as a pond contains fish) but, rather, involves other living beings as part of a well-ordered and organized functional network, which is captured, I suggest, by Leibniz's notion of domination. The notion of domination among the monads may well express this degree of functional organization.

If we try to understand the notion of a natural machine—a machine that remains a machine to the least of its parts—as a physiological characterization of *emboîtement* (of machines within machines) to infinity, we don't get very far. As Andrault observes, none of the physiological cases familiar to Leibniz, such as the structure of the heart (which he learns from Steno), or the silkworm, or the *animalcula* revealed by the microscopists, fits with Leibniz's characterization of organic bodies as machines to the least of their parts (Andrault 2014, 123–30). And, as seems fairly obvious, no empirical observation can confirm Leibniz's thesis of a nested structure that develops to infinity. Although it is clear that Leibniz draws both inspiration and support from the empirical observations of his time, the thesis itself fits with none of them. If we understand the notion of nestedness in terms of the domination relation— that is, roughly, as a degree of functional organization within an organic complex structure—it may be consistent with the monadological framework, or at least with a very significant strand of this framework.

For this reason, looking for the limit of composition or division—that is, looking for the most fundamental level by going top down from the complex to the simple— turns out to be, at least in a sense, misguided. For, as Leibniz points out to Bernoulli,

---

[17] See Andrault (2014, sec. 3.2.2).
[18] See Nachtomy (2010).

some of the most fundamental constituents of reality may be found at the top. A complete monad, as exemplified by the self, is simple in the sense of its being a unity without parts. The self has no parts in that the part/whole model does not apply to humans or animals; it only applies to aggregates or to material compositions. As Leibniz says, in order to find something that is a substance, we need not proceed down, for such things present themselves immediately, since each animal is such a thing.

We may of course proceed down the line of domination relations (see *Monadologie* 70 again), but we need not do so in order to reach the most basic unity, for every animal and every human being is such a unity. Leibniz does not provide us with better examples of this than his frequent use of the I, the ego, the whole animal, and so on. This point comes out clearly in a letter to Sophie (June 12, 1700):

> [t]hus, Unities, although they are all indefectible, are not all equally noble, and in an organic body there can only be one dominant and principal Unity, which is its soul [*âme*]. *C'est le moi en nous*, which is well above most other souls. (GP VII:554; see Fichant 2004, 341)

In an organic body there can only be one dominant and principal unity. But there are infinitely many dominated monads that belong to and constitute an organic being. Thus, we can say that monads are found not only at the bottom and not only at the top but, in fact, all over the range in between them, as well.

As Fichant observes, this universalized picture is made manifest by the fact that monads have no proper names—they do not refer to individuals as the complete concepts of the "Discourse on Metaphysics" do; rather, their employment is universal—they apply to unities at every level of organization (Fichant 2014, 136). Fichant calls this feature the universalization of the monad, where instead of the paradigmatic examples of Alexander and Caesar of the "Discourse," we find nameless monads, so that "proper names no longer find their usual reference" (131). This suggests that the scope of the concept of monad is extended—not only to individuals at the top and not only at the bottom but also to unities that operate at all levels of a functional organization.

My main suggestion in this chapter, then, is that Leibniz supposes monads to pertain not only to the bottom level but also to the top, so that they apply to what we would identify as individual humans or animals or, more broadly, to what we normally regard as organic unities, as well. And this turns out to imply that the extension of the concept of monad applies not only at the bottom and not only at the top but also all over the range in between, by means of the domination relation.

9.4 CONCLUSION AND SOME IMPLICATIONS

This point has some important implications. Let me note here just one which I find particularly interesting. Did Leibniz intend the notion of a natural or divine machine to refer to just the body of a living being, or to the whole living being? The texts are ambiguous on this point. Leibniz sometimes says this, sometimes that.[19] In *Divine Machines*, Smith (2011) seems to hold the former. Duchesneau's (1998) position seems to have evolved from the former to the latter. If Smith and Duchesneau are right that, in the exchange with Stahl, we find a clear articulation of a new concept of life, according to which life is *only* perception, and does not involve the organic body at all, then this position is further corroborated.[20] Still, as we have seen, there are strong reasons, both textual and conceptual, to think that the notion of a natural machine must apply to the whole living being.[21]

As I noted in the beginning of this chapter, however, the role monads play at the top and the role they play at the bottom are not the same. The status of monads at the bottom level is not particularly controversial and does not require special attention here. But how are we to understand the more controversial claim that monads are at the top? There are at least two ways to go here. A stronger reading of this claim holds that the created monad (once an entelechy or substantial form is completed with primary matter) constitutes a corporeal substance. This view is most boldly stated and tenaciously argued for in Phemister (2005). Phemister argues that "the corporeal substance is the true unit, and by default, therefore, supports the view that the corporeal substance is itself a monad" (72); "Indeed, corporeal substances, as animals, fish, trees, plants, and anything that possesses its own life force, . . . are . . . the best and most accessible examples of monads" (76). However, texts in which Leibniz refers to complete monads as corporeal substances are scarce. Phemister accounts for this scarcity of texts by a close analysis of Leibniz's correspondents and the distinction between his public and private writings. She argues convincingly that,

> [i]n his unpublished writings, as for instance in his correspondence with De Volder and with Des Bosses, Leibniz is fairly relaxed about openly admitting that souls or entelechies are primitive active forces and that they exist together with primary matter or primitive passive force. The unity of active and passive primitive forces is equivalent to the unity of soul or entelechy and

---

[19] See, for example, "In Machinis naturae, seu corporibus viventium, organicis" (LSC 20) vs. "in viventibus, seu in naturae Machinis" (LSC 22).

[20] According to Smith (in correspondence), on this view, the organic body is capable only of "vegetation," which is now distinct from life.

[21] For further discussion and argumentation, see Arthur (2018).

primary matter. In the De Volder Correspondence, this unity of soul or primitive entelechy and a primitive passive force or primary matter constitutes the Leibnizian monad. (31)

At the same time, the other way to go—the other way to understand the role monads play at the top—is this. While a complete animal may not *itself* be a monad, it may nevertheless be *unified* by a monad: the dominant monad, which functions as a soul in composite living things or corporeal substances. In this way, the monad appears not just at the bottom level, metaphysically speaking, but also at the top, as that which creates and unifies the living thing or corporeal substance. And this crucial role of the monad gets lost when we focus on the role of the monad as the ultimate elements of things, as Leibniz puts it in the beginning of the *Monadologie*.[22]

I will not argue for either of these interpretations here. Instead, I will restrict myself to pointing out that Leibnizian monads figure not only at the bottom level but may also be seen as grounding the reality of the organic—and therefore the truly metaphysical—world at the top. If this is true, then monads figure all over the range between them, as well. In addition, I would like to point out that this analysis gives rise to two elegant suggestions as to how to relate these two inverse directions of composition. First, it may well be that the two directions of grounding complement each other, such that the one could not work without the other. My second (admittedly radical) conjecture for further research is this: the grounding of monads from the bottom up pertains to aggregates or composites, while the grounding of complete monads from the top down pertains to complete animals and organic beings. If this is true, then we see that each direction of grounding plays a crucial role in Leibniz's final metaphysical picture.

---

[22] I owe this formulation to Dan Garber (in correspondence).

10

LIFE AND FORCE

10.1 ORGANISM AND MECHANISM, LIFE AND FORCE

In his fifth letter to Clarke, Leibniz writes the following:

> 115. As for the motions of celestial bodies, and even the formation of plants and animals, there is nothing in them that looks like a miracle except their beginning. The organism of animals is a mechanism that supposes a divine preformation. What follows upon it is purely natural and entirely mechanical.
>
> 116. What ever is performed in the body of man and of every animal is no less mechanical than what is performed in a watch. The difference is only such as ought to be between a machine of divine invention and the workmanship of such a limited artist as man is.[1]

Especially noteworthy in these passages is this line: "The organism of animals is a mechanism that supposes divine preformation: what results is something purely natural, and completely mechanical" (GP VII:417–18).

In his book *Divine Machines*, Justin Smith makes the following insightful remark:

> This is, in a word, Leibniz's theory of divine machines: divine, because initially generable only by God directly; machines, to the extent that one need take no recourse to God's constant concourse, nor to some subordinate God-like principle within the machine, in order to obtain an adequate understanding of it.[2]

Smith's analysis is certainly admirable. His observation gives rise to a subtle point I highlight here. Smith is surely right that, for Leibniz, all the phenomena of life

---

[1] G. W. Leibniz, *G. W. Leibniz and Samuel Clarke: Correspondence*, ed. and trans. R. Ariew. (Indianapolis, IN: Hackett, 2000), 63.
[2] Smith (2011, 135–36).

can be explained mechanically. But, as Leibniz argues in the context of his physics, while every phenomenon can (and should be) explained mechanically, mechanical explanations are insufficient to account for the very reason and foundation of mechanism, entailing that metaphysical explanations must be used as well. Roughly speaking, Leibniz's typical move is to say that every phenomenon can (and ought) to be explained mechanically, but at the same time, such phenomena presuppose a substantial foundation of active forces that can only be explained in metaphysical terms. I shall argue that Leibniz's subtle position regarding the distinction between artificial machines and natural machines is similar. Both are machines and hence could be explained mechanically without invoking any mysterious principles. And yet, a divine machine, having an organism that presupposes divine preformation, turns out to have some metaphysical features (intrinsic unity, infinity, life, and teleology) that artificial machines (and nonorganic things) do not have. Such features presuppose life, seen as a metaphysical concept, as their ground.

As we have observed in chapter 6, Leibniz's theory of natural machines is developed in the "New System," where his aim is precisely to limit the pretensions of Descartes and his followers that extend the mechanistic view of nature too far. There, Leibniz attempts to reconcile ancient and modern philosophies of nature by accepting mechanical description at the level of physics and Aristotelian description at the level of metaphysics. He thus endorses a mechanical description of bodies, but strongly resists the Cartesian attempt to describe natural machines in terms of artificial ones.[3]

In his controversy with Stahl, Leibniz takes up this point and makes a very interesting remark regarding the subtle relation between organism and mechanism. He writes,

> all organism is in fact mechanism, but more exquisite and, so to speak, more divine. And it may thus be said (as I already noted), that the organic bodies of nature are in truth divine machines.[4]

---

[3] AG 141–42; GP IV:481. A similar point regarding our tendency to overlook the greatness of nature is articulated in the correspondence with Clarke: "We would like that Nature would not go further, that it would be finite, as our spirit: but this is not to know (appreciate) the greatness [*grandeur*] and majesty of the Author of things. The least corpuscle is actually subdivided to infinity, and contains a world of new creatures, of which the universe would lack if this corpuscle would be an Atom, that is to say, a body of one piece without subdivision" (letter to Clarke; GP VII:377).

[4] *Animadversion* 2; LSC 31.

In the introduction to their new edition of the Leibniz/Stahl controversy (LSC), Duchesneau and Smith comment on Leibniz's view:

> in *Animadversion* 2, Leibniz opts for identifying organism with a special case of mechanism: organism is mechanism deploying itself in organic bodies understood as machines of nature. Resulting from divine preformation, machines of nature display their inner sufficient reasons to infinity by their organic dispositions. This is to be understood in two ways: there is no mass, however small, that does not contain within itself organic bodies or machines of nature, and thus there is an indefinite unfolding of organicity in the analysis of natural phenomena, since they all display inherent teleology; but machines of nature have ends that follow from the "force of their structure"—they have been internally contrived to display functional effects of their own—and not simply from series of multiple concurring external conditions (Ex1). But although they illustrate God's supreme craftsmanship, they abide by the regular laws of physical nature, which are expressive of and rule over mechanism (Ex2).[5]

Again, I fully agree with Duchesneau and Smith here. The main point I would like to highlight is this. Just as Leibniz invokes force to account for the phenomena of motion in physics, so in his life sciences he invokes the soul (or *anima* or entelechy) that must be presupposed as *grounds* for the phenomena of life. Indeed, Leibniz's motivation to invoke the notion of a natural machine is precisely to limit the extension of mechanical philosophy too much and to draw a line between living and nonliving things. But as already suggested in chapter 6, this distinction does not turn primarily on physiological grounds but, rather, involves metaphysical considerations.

Thus, I think that there is a sense in which Leibniz does use a God-like principle within the machine in order to obtain an adequate understanding of it; that principle is life itself or, more precisely, its source. But as with the notion of force, the source of life, I suggest, must be understood in a *metaphysical* sense.[6] In his dynamics, Leibniz has recourse to the soul, a principle of action, or to something analogous, seen as a metaphysical principle; his attitude with respect to life, in the context of his life sciences or his account of living beings, is, I suggest, similar. In both cases, these metaphysical principles are supposed to be entirely compatible with a mechanical explanation of the phenomena of life.[7]

---

[5] LSC lv, lvi.
[6] "All of these operations are to be explained separately once the reason of the first motion is understood, or, which is the same, that of Life" (Smith 2011, app. 2, 289).
[7] See paragraph 124 in Leibniz's fifth letter to Clarke (in Leibniz 2000, 46).

10.2 LIFE AND FORCE—METAPHYSICAL CONCEPTS

One of the constant and most fundamental features of Leibniz's metaphysics is that a substance or anything that truly exists is informed by an internal source of action—an entelechy, an *âme*, a soul, a principle of action, *conatus*, primitive force, or a principle of life, to use only some of the terms Leibniz employs, sometimes utterly interchangeably (e.g., *Monadologie* 66, 70, 74; Theodicy §§396, 360–61). Evidently, there are several terms and concepts that serve Leibniz to make this point—terms that relate to various historical sources and serve different polemical purposes. It goes without saying that there are significant differences between these terms and concepts, but for my purposes here, those differences can be ignored.[8] For the question I would like to raise here is whether Leibniz's use of the notion of life is analogous to his use of the notion of primitive force in his critique of a purely mechanical view of nature. Obviously, Leibniz is very keen to show that not only are there living things everywhere in nature, such that there is nothing sterile or lifeless in the universe (e.g., *Monadologie* 66), but also that the very foundation of nature consists in living things, and that without them there would be nothing at all.

We know that, for Leibniz, the notion of primitive force is a defining feature of substance; we also know that the notion of life comes to play a similar role, so that it pertains only to living beings to have such power or force. If any living thing has an internal source of action and everything with an internal source of action is living, one wonders what the difference between Leibniz's use of the notion of life and his use of the notion of force (or entelechy or appetite, for that matter) is.[9] One can put this question as follows: If any animate being has an anima (or is soul-like), and any thing that has an anima is animate, and if we know that any true being for Leibniz has a source of activity (entelechy or soul-like thing), is there a significant difference between a source of life and a source of being? A quick response to this question is that there is no substantial difference, and this is why, at the metaphysical level, primitive force or a source of life is considered as the ground for all beings. Differences arise, however, when these metaphysical principles are used in the context of physics and the more particular life sciences.

---

[8] Leibniz himself makes the point that the terminology in this case is of no importance to him. In the "Specimen of Dynamics," he writes, "we must admit something metaphysical, something perceptible by the mind alone over and above that which is purely mathematical and object to the imagination. . . . Whether we call this principle form or entelechy or force does not matter, as long as we remember that it can only be explained through the notion of forces" (AG 125).

[9] A related question is what role (if any) the notion of life that Leibniz uses in his metaphysics plays in his writings on particular life sciences. But I cannot explore this further issue in the space available here.

To illustrate how Leibniz uses the notion of entelechy or primitive motive force in the context of his critique of a purely mechanistic conception of nature, consider the following passage from "On Nature Itself":

> we must judge . . . that a first entelechy must be found in corporeal substance, a first subject of activity, namely a primitive motive force which, added over and above to extension (or that which is merely geometrical), and over and above bulk (or that which is merely material), always acts but yet is modified in various ways in the collisions of bodies through *conatus* and *impetus*. And this substantial principle itself is what is called *the soul* in living things and *the substantial form* in other things; insofar as, together with matter, it constitutes a substance that is truly one, or something one *per se*, it makes up what I call a monad.[10]

As I argued in the previous chapter, monads figure not only at the bottom level, accounting for the composition of aggregates, but also at the top, accounting for the unity of animals. My motivation for comparing the notions of life and force here is mainly to argue that both should be seen as metaphysical concepts.[11] The point of making this claim will become clearer in the next section. For now, let us attend to what Leibniz writes to Sophie Charlotte:

> I am even inclined to think that souls or active forms are found everywhere. And in constituting a complete substance, matter cannot do without them, since force and action are found everywhere, and since the laws of force depend upon

---

[10] AG 162; GP IV:511. See also the following passage from "On Nature Itself": "investigating the matter with greater care, we must distinguish the ultimate sources [*principia*] in this mechanism from that which is derived from them. . . . [T]he source of the mechanism itself flows not from material principles and mathematical reasons alone, but from a higher and, so to speak, metaphysical source, something that I think will be of use in preventing the mechanical explanation of natural things from being extended too far, to the detriment of piety, as if matter can stand by itself and as if mechanism required no intelligence or spiritual substance" (AG 156–57; GP IV:504–505).

[11] Rutherford ("Leibniz on Infinitesimals and the Reality of Force," in *Infinitesimal Differences: Controversies between Leibniz and his Contemporaries,* ed. U. Goldenbaum and D. Jesseph [Berlin and New York: Walter de Gruyter, 2008], 255–80) provides very convincing arguments for seeing Leibniz's notion of force as metaphysical by considering the relations between force and Leibniz's notion of the infinitesimal. Rutherford formulates the following dilemma: "if force is real, it cannot be an infinitesimal quantity; if it is an infinitesimal quantity, it cannot be real" (256). He concludes that "[f]orce, instead [of being mathematical or physical] is something 'metaphysical,' which is not representable by the imagination, but can be grasped only by the intellect" (272).

some marvelous principles of metaphysics or upon intelligible notions, and cannot be explained by material notions or the notions of mathematics alone or by those falling under the jurisdiction of imagination. (GP VI:507; AG 192)

Saying that the notions of force and life are metaphysical is meant to stress that they cannot be fully understood in physical or mathematical terms. Force and life cannot be described because they are not part of an empirical description of nature; rather, they constitute its foundation, which cannot be captured by empirical science or mathematics.[12] This would also highlight something about which Leibniz is quite explicit in his dynamics—that is, the intermediate status of force as something that, on the one hand, cannot be reduced to mechanistic terms and, on the other, is partly expressible in mathematical terms. Indeed, the famous formula of the *vis viva*, or living force—mv^2—points to the intimate connection that Leibniz presupposes between force and life (via the notion of activity).[13] I do not mean to place too much weight here on the term *living* force, though it is certainly suggestive; the dead/living force terminology is present early in Leibniz's work (1674) and derives from Galileo.[14] But I do want to stress that both force and life presuppose a source of action, as indicated at the beginning of this section.[15]

10.3 "THESE PRINCIPLES OF LIFE OR THESE SOULS HAVE PERCEPTION AND APPETITE"

Let us now consider the connection between force and life in the context of Leibniz's debate over Ralph Cudworth's plastic natures and Nehemiah Grew's principles of life. In this context, Leibniz writes:

> I also suppose everywhere these principles of perception and life that the Cartesians accord only to a small part of matter by a rather odd exception that is far removed from nature's usage.[16]

---

[12] See also GP IV:465.
[13] This status may indicate that force, like the intermediate degree of infinity, has both quantitative and qualitative aspects.
[14] I thank Richard Arthur for this observation.
[15] It is also remarkable that Leibniz's adherence to the conservation of this force forms an important part of the background to his debate with Clarke. For the conservation of this force in living beings is precisely what allows Leibniz to argue, contra Clarke, that God need not intervene with the regular course of nature, for the same amount of force always persists in creatures.
[16] "Je suppose aussi partout ces principes de perception et de vie que les Cartésiens n'accordent qu'à une petite partie de la matière, par une exception tout à fait étrange et bien éloignée de l'usage de la nature" (*Éclaircissements*; GP VI:551).

The principles of life belong only to organic bodies. It is true (according to my System) that there is no portion of matter in which there would be no infinity of organic and animate bodies, by which I understand not only animals and plants, but perhaps also other sorts entirely unknown to us. But one should not say on this account that each portion of matter is animate, just as we do not say that a pond full of fish is an animate body, even though the [individual] fish is.[17]

In his late writings, Leibniz's notion of life (*vie*, in the passage just quoted) seems to be limited to perception and appetition. He makes this point explicitly in his dispute with Stahl as well: "I am in the habit of locating life in perception and appetite";[18] "If the body should be devoid of perception and appetition, I believe that it would no more deserve to be called living more than a flame that labors to feed itself."[19] As Duchesneau and Smith emphasize in their introduction to the *Leibniz-Stahl Controversy*, in this text, we find him equating life with the monads' capacity for perception (LSC lxi).[20] Indeed, Leibniz vacillates on the precise meaning he ascribes to life. I think that some of this apparent vacillation is due to the use of "life" in two closely related contexts: one in which the phenomena of life are emphasized and another in which the source (or principle) of life is emphasized.

Indeed, in the same text, Leibniz writes, "it is most true that the soul, equally suited to this work by divine preformation, acts by means of its perception and appetite."[21] It is quite clear, however, that Leibniz's notion of *appétit* is strongly related to his notion of power of action or primitive force, which he ascribes to the soul, for it seems to be nothing over and above the drive—or the conatus—to move from one perception to another.[22] This seems analogous to Leibniz's distinction between the

---

[17] "Les principes de Vie n'appartiennent qu'aux corps organiques. Il est vrai (selon mon Système) qu'il n'y a point de portion de la matière, où il n'y ait une infinité de corps organiques et animés; sous lesquels je comprends non seulement les animaux et les plantes, mais encore d'autres sortes peut-être, qui nous sont entièrement inconnues. Mais il ne faut point dire pour cela, que chaque portion de la matière est animée, c'est comme nous ne disons pas qu'un étang plein de poissons est un corps animé, quoique le poisson le soit" (GP VI:539).
[18] LSC 33–35.
[19] LSC 277.
[20] Duchesneau and Smith ascribe this capacity of perception to a non-bodily feature of the monads. But, as I argued in the previous chapter, monads at the top come united with bodies.
[21] *Animadversion* 8; LSC 27.
[22] As Julia Jorati (2017, 36) recently put this point: "The changing states of a monad, which Leibniz views as modifications of the monad's primitive force, are that monad's perceptions and appetitions." This topic has also drawn some attention at the recent International

power of action and action itself. In each created substance or living being, an intrinsic power of action or force is required for the execution of particular actions and is presupposed by the transition from one to the other. As he writes,

> active force contains a certain act or entelechy, and is something of a middle nature between the faculty of acting and act itself; it involves a *conatus* or endeavor, and is of itself carried into action and stands in need of no help, but only that the impediment is taken away.[23]

Thus it seems that, at least in his later writings, Leibniz tends to capture the notion of action in terms of perception and the notion of power of action in terms of appetition.[24] Another example in which the notions of force, perception, and life come together appears in Leibniz's letter to Sophie from June 12, 1700:

> Since this concerns Unities of substance, it is necessary that there will be force and perception in these very unities. . . . It is required therefore that, in bodies that have sentiments [sentient bodies], there would be *unique substances, or Unities, that would have perception, and it is this simple substance*, this Unity of substance or this Monade, that one calls the soul [*âme*]; and by consequence, souls, as any other Unity of substance, are immaterial, indivisible, imperishable. . . . And once these unities are given life [*ont une fois de la vie*], it is necessary that they will be immortal and live forever. These Unities truly constitute substances and each Unity makes uniquely one substance.[25]

This passage highlights the connection between the notions of unity, force, and perception. In addition, it seems that instead of saying that "once these unities are given life," Leibniz could well have said that once these unities are given force or power of action, they will never cease to act and thus would be immortal and live forever (unless annihilated). This very formulation comes up in "On Nature Itself," for example. This point is made even stronger and clearer when we recall that the only context in which substances are ever given life or force is the very context of creation. And

---

Leibniz Congress (X). See, for example, Stephan Schmid, "The Intrinsic Directedness of Leibnizian Forces," in *Vorträge des X. Internationalen Leibniz-Kongresses*, ed. Wenchao Li (Hildesheim: Georg Olms, 2016), 5:131–41.
[23] "On the Correction of Metaphysics and the Concept of Substance" (L 433).
[24] It is arguable that this view is articulated much earlier. In his recent book (2018), Richard Arthur argues that it arose in the late 1670s. What I emphasize here is that there is certainly a shift in terminology toward appetition and perception.
[25] GP VII:552–53.

this makes it rather clear that there would not be any significant difference between endowing a creature with life, primitive force, or power to act; for, in the context of Leibniz's metaphysics, being endowed with life or primitive force is synonymous with being created.[26]

10.4 LIFE AND THE PHENOMENA OF LIFE

For this reason, let us pursue the analogy between life and force a bit further. This point is subtle; it concerns the relations between attributing life to living beings and the phenomena of life—that is, the phenomena that are traditionally taken to characterize living beings, such as locomotion, sensation, self-nourishment, generation, sensation, and so on. In her recent book, Raphaele Andrault argues convincingly that there is a certain discrepancy in Leibniz between the phenomena of life and life itself.[27] She explains this as follows:

> Because life is not merely an appearance, is not a phenomenon but the foundation of any phenomenon, it is precisely impossible, on the basis of corporeal phenomena, to determine what is vital and what is not. In themselves, the phenomena are not vital, or are so only metonymically, but their foundations are all vital.[28]

In his "New Essays" (NE), Leibniz suggests that, if something appears to be living, such that it manifests phenomena similar to those of living beings, it does not follow that it is living. For, "otherwise it will be only an appearance, as the life that American savages have attributed to clocks and horologes, or to marionettes . . . which they believe to be animated by demons."[29]

Life presents observable appearances typical to living things, but it cannot be reduced to (or fully expressed in) terms of such appearances; rather, it is something that needs to be presupposed as the basis or the foundation of living phenomena. This is why one cannot infer from observing something that appears to be living that it is living. In its Leibnizian sense, life is understood as a metaphysical notion; it is seen as an internal power of action or force. As such, it is not the kind of thing that can

---

[26] For more details, see Nachtomy (2007a, ch. 5).
[27] Andrault (2014).
[28] "Parce que la vie n'est pas qu'une apparence, n'est pas un phénomène mais le fondement de tout phénomène, il est précisément impossible de déterminer à partir des phénomènes corporels ceux qui sont vitaux et ceux qui ne le sont pas. Les phénomènes en eux-mêmes ne sont pas vitaux, ou seulement par métonymie, et leurs fondements le sont tous" (Andrault 2014, 204).
[29] A 6.6:348–49.

be observed only its manifestations can be observed. For Leibniz, I suggest, life and force are seen as metaphysical, not physical, concepts. And this is why they cannot be directly observed, but can only be understood (or made intelligible) by the intellect. This point, too, is strongly related to Leibniz's position in the correspondence with Clarke. As Maria Rosa Antognazza nicely puts this, once created, "the physical autonomy of the world mechanism does not imply its metaphysical independence of God. . . . Leibniz shifted the dependence of the universe on God from the physical to the metaphysical level."[30] It is in this sense that a metaphysical God-like principle is presupposed within natural machines. That principle, I suggest, is what Leibniz regards as life itself. For this reason, life can be taken to belong to the soul alone. But this would be inadequate, I think. For what Leibniz regards as living is not the soul alone but, rather, the whole individual or animal whose soul it is.

10.5 METAPHYSICS AND THE LIFE SCIENCES

This also explains why there need not be a direct relation between life as a metaphysical concept and the empirical observations that Leibniz makes in the particular life sciences. More precisely, this is why Leibniz holds—which is something we can gather from his practice—that in some sense the metaphysics of life and the phenomena of life are almost independent of one another. In other words, the notions of life and force are fundamental in metaphysics, and they must be presupposed for explaining motion and the phenomena of life; but for this very reason, they need not be used for explaining the details of both the physical and the biological worlds. This point is developed a little further in the present section.

Leibniz's definition of created substances as animate, infinite beings has significant consequences in his metaphysics. This definition is particularly pertinent in regard to living beings as natural and divine machines—that is, machines that remain machines in the least of their parts. In seeing created beings as living, Leibniz seeks to introduce teleology into the natural world and thus to avoid the consequences of a purely mechanistic picture of nature; the definition of created beings as infinite also serves to emphasize their likeness to their creator; similarly, it serves their definition as basic unities, since they are united by a single law, informing their development and regulating their changes of state. Such metaphysical roles are crucial to Leibniz's overall agenda and philosophy. By developing the analogy that Leibniz presupposes between infinity and perfection (see section 8.2), I showed how Leibniz's use of infinity serves his attempt to introduce moral value into created things and, by extension, into the world as a whole. As we have seen in his critique of Pascal (chapter 7),

---

[30] Antognazza (2009, 535).

Leibniz offers a vision of the living world in which intrinsic connectedness and harmony reign, in contrast to Pascal, who depicts the human spirit as lost (*égaré*) between the infinitely large and the infinitely small.

While the metaphysical and theological significance of these points is rather clear, questions arise about the empirical consequences of defining created beings as living and infinite. In other words, we have seen that life involves a kind of infinity and a kind of force, but what does that tell us about actual living creatures? In reading some of Leibniz's life-scientific texts, one wonders what the consequences of Leibniz's metaphysics of living beings are for his engagement with the particular life sciences of his time (physiology, medicine, embryology, microscopy, etc.).[31]

One strategy to approach this question is to argue that Leibniz runs his metaphysical and life-scientific investigations side by side for different purposes, as a methodological (nonexclusive) pluralism. Indeed, I believe that this is the most promising supposition to take in light of relations between Leibniz's metaphysical positions and the empirical observations he endorses in his medical/scientific texts. But this hypothesis should be further reassessed, especially in light of the texts that are about to appear, which remains a project for future work.[32]

Here I will consider only one example, viz., Leibniz's recommendations regarding classification methodology in his letter to Gackenholtz of 1701.[33] While he indeed mentions his early work on the combinatorial art as a general background, he in fact warns against making too much of a priori categories and recommends paying close attention to observation. While he regards each organic being as a machine of nature (with a certain function), his advice on classification seems to draw much on the peculiarities of biological observations.[34] For instance, Leibniz holds that focusing on one set of properties or criteria would not suffice to classify plants into species; rather, as Justin Smith points out, "a biological species is for Leibniz a relatively isolated reproductive community, all of whose members may be said to be members of the same species not in virtue of any morphological resemblances (though they generally have these as well) but rather in virtue of shared origins" (2011, 236).

While Leibniz's metaphysics commits him to a rigid position regarding a fixed set of individuals and within them a fixed set of properties (expressed by their complete concepts), his view of classification is surprisingly liberal and pragmatic. As it turns

---

[31] This question becomes especially pertinent in light of newly edited and translated materials from Leibniz's corpus of medical writings.
[32] The present articulation of this strategy has been suggested to me by Barnaby Hutchins.
[33] This has been translated and published by Smith as "On Botanical Method," in Smith (2011, app. 5).
[34] "Plants, animals, and, if I may say it in a word, organic bodies that are produced by nature, are machines fitted for the perpetuation of certain functions" (Smith 2011, app. 5, 305).

out, he is observing that classification depends on a certain criterion for grouping individuals into classes—and in his mind, there is no privileged criterion for classification (see Smith 2011, 244). Thus, for example, in the letter to Gackenholtz, he is arguing against the tendency of contemporary botanists to privilege flower morphology as *the* classification criterion. As he writes,

> it is necessary that plants, as you well remind us, be distinguished according to features other than the flowers, which are not always visible. I mentioned the root to you as an example. . . . Further, the structure of the whole plant should not be neglected either, from which first arises the division into trees, bushes, and grasses. Coming to the parts of the plant, what a multitude of criteria for division are to be found there, either in the solid parts, as in the roots (of which I have already spoken to you), the trunk (its interior and its bark), the leaves (either of the plant or of the flower), the fruit (in which the seeds are discovered), and in other solid parts; or in the fluid parts, such as the marrow, the water that flows out, the resin, and, finally, in the humid substance that is pressed out, or, more precisely, in the odor or vapor that is emitted; meanwhile different plants are chiefly distinguished by different parts that they contain or to which they give rise; some by the flower, others by the root, others by the bark, some by the sap, etc. And there is none of those parts from which a certain comparison could not be drawn with wide application throughout the vegetable kingdom. (cited in Smith 2011, app. 5, 307)

Leibniz asks,

> [h]ow many are the species of reeds, in which the stalk predominates? Why do we not group together the plants that yield berries in the same way that we do the grain-yielding ones? Why not in the same way the whole genus of ferns? (307–308)

In contrast with jurisprudence,

> it is more relevant concerning nature that it be considered in all aspects, that all manners of comparison be instituted, since the investigation of it is more difficult than that of civil matters. (308)

Leibniz's attitude here reveals a combination of good practical judgment and sensitivity to careful observation and to fine detail. In addition, it also reveals his willingness to apply different methods in different domains of investigation (here, jurisprudence and plant classification)—an attitude familiar from his many other interests, such as geology and medicine, among others. In short, his attitude here exemplifies well his methodological pluralism.

As discussed in chapter 9, Leibniz's characterization of natural machines as machines within machines to infinity does not fit with his empirical observations—and, indeed, cannot ever do so. At the same time, understating Leibniz's thesis of a natural machine as one that distinguishes between living and nonliving (animate and inanimate) things turns out to be quite insightful—but not so much for its empirical import. If this thesis is understood as characterizing organic beings as systems of functional relations, such that each constituent both serves the whole animal (or organism, in the contemporary sense of the word) and is at the same time served by other constituents, while the whole thing is considered a single system, unified by certain rules of development, it does seem an insightful way to draw a line between living and nonliving things. Putting this into a broad historical context, we see that this notion of a natural machine could have contributed to Kant's influential definition of an organism, which in turn should be seen against the central role the notion of organism has come to play in the new science of biology. It is perhaps not superfluous to remind the reader that the term "organism" was more or less invented by Leibniz at the turn of the seventeenth century, and that the term did not refer to single organic unities at the time.

I say all this in order to stress that, as long as we understand the notion of a natural machine in a metaphysical sense—one that seeks to clarify the essence of animate beings, and one that may inspire (but also need not get in the way of) empirical investigations—both domains, that of metaphysics and that of the particular life sciences, might be well served.

It is in this light that we need to read Leibniz's dictum that "an organism is nothing but a mechanism but more exquisite and divine." Organic bodies (or the organism of animals) should be studied like any mechanism—that is, by the methods appropriate to the life sciences (observation, quantification, classification, etc.)—but as in physics, one need not neglect the origin and the higher principles underlying it (that is, principles concerning force). In the life sciences, one should not neglect the source of the activities of living beings—that is, life itself or, better, its source. Like primitive force, for Leibniz, life is seen as a divine gift. Like force, it is not something that comes and goes; rather, life is constitutive of creatures and, by consequence, of the natural world. For this very reason, life and force are not the proper objects of the empirical life sciences but, rather, of metaphysical reflection. (It is also in this sense that Leibniz's notion of primitive force may be regarded as infinite in kind.)

CONCLUSION
THE RE-ENCHANTMENT OF NATURE

In conclusion, let us attend to Leibniz's broad aims and motivation. I think that the best way to see this is to recall once more the historical context. What I have in mind is Leibniz's subtle response to the Cartesian attempt to mechanize both nature in general and the domain of living things in particular. Descartes attempted to mechanize virtually all the functions that had traditionally been assigned to the vegetative and sensitive souls. His vision of nature, which made it particularly amenable to mechanization, consists of bits of matter in motion: matter is devoid of any powers, activity, or life, and thus involves mere extension. In this sense, Descartes attacked the ancient view of nature and replaced it with a thoroughly disenchanted view of nature.

This program had significant consequences. On the one hand, it led to a radicalization of the mechanistic agenda, in which, for some thinkers subsequent to Descartes, even the mental becomes naturalized, thus eliminating Descartes's problematic mind/body dualism. This approach is evident in the work of eighteenth-century physiologists, and was very clearly expressed in La Mettrie's *L'Homme Machine* (1748). On the other hand, Descartes's agenda faced strong resistance. It led to the invocation of various kinds of incorporeal principles, vital forces, and "plastic natures" (e.g., More, Cudworth, and Grew), which were supposed to be in living beings and to be irreducible to mechanization and resistant to description in quantitative terms. This strand might be called vitalism or animism.

Avoiding these two extreme positions, Leibniz offers a subtle and influential—if underappreciated—reconciliation. Leibniz, whose talents for conciliating opposing views by producing an original synthesis are well known, had worked out an ingenious way to bypass the debate between vitalists and mechanists. Leibniz developed a sophisticated way to fully embrace mechanism while avoiding its more dire consequences, by reintroducing life into the very foundation of the natural world. For Leibniz, "[m]atter [itself] is plastic or organic throughout, even in those portions that are as small as can be conceived" (GP III:368). As he notes in his comment on

Pascal, "[w]hat wouldn't he [Pascal] have said with his powerful eloquence if he had gone further, if he had known that all matter is organic?"[1] Or, as he puts it in the "Specimen of Dynamics," "I admit an active and, so to speak, vital principle superior to the common notion of matter everywhere in bodies" (AG 125). Leibniz's view of matter itself as organic, such that the very foundation of nature consists of living beings, makes the need to invoke some extra vital forces or plastic natures (of whatever sort) redundant. Most important, the organic nature of matter comes with built-in activity, teleology, and life—but without animism as such, in that it involves nothing added on top of matter. In brief, this move is what I am calling Leibniz's "re-enchantment of nature."

As a part of this re-enchantment, Leibniz sought to resist the "pretensions" of the sweeping application of the (Cartesian and post-Cartesian) mechanistic understanding of nature. Leibniz worried that Descartes's agenda would lead to a view of nature devoid of value. This, he aptly calls naturalism: "Spinoza begins where Descartes leaves off: *in naturalism*" (AG 277). Such a naturalism would, for Leibniz, lead to Spinoza's "dangerous views" (expressed, for example, in "Two Sects of Naturalists," AG 282). On the metaphysical/theological level, Leibniz's response to these Cartesian–Spinozist threats was to propose a metaphysical picture in which other worlds are possible, so that the actual world is selected for its goodness, because it is the best of all possible worlds. This use of logical possibilities and a reasoned choice among them would make room both for value (or goodness) in the world, as well for free and reasoned choice between possibilities.

On the level of natural philosophy, Leibniz sought to ground the very ontology of the natural world in divinely created, living machines endowed with an internal law and primitive force. In his "New System," he explicitly introduces the notion of a natural machine (in distinction to an artificial machine), in order to limit the danger of all-inclusive mechanism. In particular, Leibniz seeks to resist (and reject) Descartes's reduction of animals to mere machines, which differ from artificial machines only in degree of intricacy and complexity, but not in kind. By contrast, Leibniz argues that God-created machines that populate the living world, such as a hibiscus plant or any common dog, are of an entirely different sort from machines produced by human artifice.

As we have seen in chapter 6, Leibniz's definition of a natural machine (as involving infinity) is of great import. A natural machine is defined as a machine that "remains a machine to the least of its parts," which implies a nested structure of machines within machines that develops *ad infinitum*. As Leibniz states in his controversy with Stahl, "all organism is in fact mechanism, but more exquisite and, so to speak, more divine.

---

[1] Buzon (2010b, 554).

And it may thus be said (as I already noted), that the organic bodies of nature are in truth divine machines."² While Leibniz's notion of a natural machine has been the object of some recent studies, his definition of life in mechanistic terms remains to be fully appreciated. His dictum (cited here) that "organism is formally nothing other than mechanism, only more exquisite and divine" needs to be understood in terms of what he seeks to achieve with it—that is, the rejection of a dead and morally neutral vision of nature through the infusion of life into its very foundation.

Describing a mechanism is one thing; understanding the reason or the point of a mechanism is another. While the former is an object of empirical science, the latter is an object proper to metaphysics (as I argued in section 10.4). As Leibniz notes in a piece from 1677, "the description of motion in a mill-house is one thing, while the description of its various applications for extracting oil, crushing grain, splitting timber, which may be brought about by the work of this mill, is another." He then goes on to make a rather surprising claim: "All of these operations are to be explained separately once the reason of the first motion is understood, or, which is the same, that of Life" ("The Animal Machine," cited in Smith 2011, app. 2, 289).³ As we have seen, the source of life in created things testifies to the presence of a divine-like agency or inherent force—a primitive force that, while it always remains the same, cannot be reduced to mechanistic terms, and that is the proper business of metaphysics.

Throughout this book, I have stressed that the infusion of life for Leibniz comes with inherent unity, agency, and infinity as metaphysical features of the fundamental constituents of the natural world. All this, I hope, highlights Leibniz's ingenious attempt to re-enchant a world that Cartesian philosophy had left disenchanted.

---

[2] *Animadversion* 2; LSC 31.
[3] Andrault goes as far as to argue that the best expression of the inner vital principle, the entelechy, is not biological phenomena but rather mechanical phenomena, such as motion (2014, 190).

BIBLIOGRAPHY

Primary Sources

Aquinas. *Summa Theologiae: Questions on God*. Edited by B. Davies and B. Leftow. Cambridge: Cambridge University Press, 2006.

Aristotle. *The Complete Works of Aristotle*. Edited by J. Barnes, translated by A. Platt. Princeton, NJ: Princeton University Press, 1984.

Galileo, G. *Dialogues Concerning Two New Sciences*. Translated by H. Crew and A. de Salvio, with an introduction by A. Favaro. New York: Dover, 1914.

Hobbes, T. *The English Works of Thomas Hobbes of Malmesbury*. 11 vols. Edited by Sir W. Molesworth. London: Bohn, 1839–1845; reprinted 1966.

Kant, I. *Gesammelte Schriften*, German ed. *Königlich Preussische Akademie der Wissenschaften*. 29 vols. Berlin: Walter de Gruyter, 1902–.

Kant, I. *Critique of the Power of Judgment. Cambridge Edition of the Works of Immanuel Kant*. Edited by P. Guyer and A. Wood. Cambridge: Cambridge University Press, 2000.

Leibniz, G. W. *Gottfried Wilhelm Leibniz: Philosophische Schriften*. Edited and translated by H. H. Holz. 5 vols. Darmstadt: Wissenschaftliche Buchgesselschaft, 1985.

Leibniz, G. W. *Recherches générales sur l'analyse des notions et des vérités, 24 thèses métaphysiques et autres textes logiques et métaphysiques*. Edited and translated by J. B. Rauzy. Paris: Presses Universitaires de France, 1995.

Leibniz, G. W. *G.W. Leibniz and Samuel Clarke: Correspondence*. Edited and translated by R. Ariew. Indianapolis, IN: Hackett, 2000.

Leibniz, G. W. *The Shorter Leibniz Texts*. Edited and translated by L. Strickland. London and New York: Continuum, 2006.

Leibniz, G. W. *Leibniz-Translations.com*. Translated by L. Strickland, n.d. http://www.leibniz-translations.com/pascal.htm.

Malebranche, N. *Oeuvres complètes de Malebranche*. Edited by A. Robinet. Paris: J. Vrin, 1958–1984.

Pascal, B. *Pensées*. In *Œuvres Complètes*, edited by L. Lafuma. Paris: Éditions du Seuil, 1963.

Pascal, B. *L'esprit de la Géométrie*. Edited by B. Clerté and M. Lhoste-Navarre. Paris: Bordas, 1986.

Secondary Sources

Adams, R. M. *Leibniz: Determinist, Theist, Idealist*. New York: Oxford University Press, 1994.

Agostini, I. *L'infinità di Dio*. Rome: Editori Riuniti University Press, 2008.

Aiton, E. *Leibniz: A Biography*. Bristol: Adam Hilger, 1985.

Anapolitanos, D. A. *Leibniz: Representation, Continuity and the Spatiotemporal*. Dordrecht: Klower Academic, 1999.

Andrault, R. *La vie selon la raison. Physiologie et métaphysique chez Spinoza et Leibniz.* Paris: Champion, 2014.

Antognazza, M. R. "Leibniz *de Deo Trino*: Philosophical Aspects of Leibniz's Conception of the Trinity." *Religious Studies* 37 (2001): 1–13.

Antognazza, M. R. *Leibniz: An Intellectual Biography.* Cambridge: Cambridge University Press, 2009.

Antognazza, M. R. "The Hypercategorematic Infinite." *Leibniz Review* 25 (2015): 5–30.

Antognazza, M. R. "God, Creatures, and Neoplatonism in Leibniz." In *Vorträge des X. Internationalen Leibniz-Kongresses*, edited by W. Li, 3:351–64. Hildesheim: Georg Olms, 2016.

Ariew, R. "The Infinite in Spinoza's Philosophy." In *Spinoza: Issues and Directions.* Proceedings of the Chicago Spinoza Conference, September 1986. Vol. 14, edited by E. Curley and P. F. Moreau, 16–31. Studies in Intellectual History. Leiden and New York: Brill, 1990.

Arthur, R. T. W. "Leibniz on Continuity." In *PSA: Proceedings of the Biennial Meeting of the Philosophy of Science Association*, 107–15. Pittsburgh: Philosophy of Science Association, 1986.

Arthur, R. T. W. "Leibniz on Infinite Number, Infinite Wholes and the Whole World: A Reply to Gregory Brown." *Leibniz Review* 11 (2001): 102–16.

Arthur, R. T. W. "Leery Bedfellows: Newton and Leibniz on the Status of Infinitesimals." In *Infinitesimal Differences: Controversies Between Leibniz and His Contemporaries*, edited by U. Goldenbaum and D. Jesseph, 7–30. Berlin and New York: Walter de Gruyter, 2008.

Arthur, R. T. W. "Actual Infinitesimals in Leibniz's Early Thought." In *The Philosophy of the Young Leibniz, Studia Leibnitiana Sonderheft*, edited by M. Kulstad, M. Lærke, and D. Snyder, 11–28. Stuttgart: Franz Steiner Verlag, 2009.

Arthur, R. T. W. "Presupposition, Aggregation, and Leibniz's Argument for a Plurality of Substances." *Leibniz Review* 21 (2011): 91–115.

Arthur, R. T. W. "Leibniz's Syncategorematic Infinitesimals, Smooth Infinitesimal Analysis, and Second Order Differentials." *Archive for History of Exact Sciences* 67 (2013): 553–93.

Arthur, R. T. W. *Leibniz.* Cambridge: Polity Press, 2014.

Arthur, R. T. W. *Monads, Composition, and Force: Ariadnean Threads Through Leibniz's Labyrinth.* Oxford: Oxford University Press, 2018.

Arthur, R. T. W., and D. Rabouin. "Leibniz's Infinitesimals: A Clarification and Defence of the Syncategorematic Interpretation." In progress.

Baruzi, J. *Leibniz et l'organisation religieuse de la terre.* Paris: Félix Alcan, 1907.

Baruzi, J. *Leibniz.* Paris: Bloud, 1909.

Bassler, O. B. "Leibniz on the Indefinite as Infinite." *Review of Metaphysics* 51 (1998): 849–79.

Beeley, P. "Approaching Infinity: Philosophical Consequences of Leibniz's Mathematical Investigations." In *Paris and Thereafter: The Philosophy of the Young Leibniz*, edited by M. Kulsad, M. Lærke, and D. Snyder, 29–48. (Studia Leibnitiana Supplamenta: Franz Steiner Verlag, 2009.

Belaval, Y. *Leibniz critique de Descartes.* Paris: Gallimard, 1960.

Bosinelli, F. C. M. "Über Leibniz' Unendlichkeitstheory." *Studia Leibnitiana* 23, no. 2 (1991): 151–69.

Bosinelli, F., and A. Lamarra, eds. *L'infinito in Leibniz Problemi e terminologia.* Rome: G.W. Leibniz-Gesellschaft, Niedersachsische Landesbibliothek, 1990.

Brown, G. "Leibniz on Wholes, Unities, and Infinite Number." *Leibniz Review* 10 (2000): 21–51.

Brown, G. "Leibniz's Mathematical Argument Against a Soul of the World." *British Journal for the History of Philosophy* 13, no. 3 (2005): 449–88.

Buzon, F., de. "Que Lire Dans Les Deux Infinis ? Remarques Sur Une Lecture Leibnizienne." *Les Études Philosophiques* 4, no. 1 (2010a): 535–48.

Buzon, F., de. "Double Infinité Chez Pascal Et Monade. Essai de Reconstitution Des Deux États Du Texte." *Les Études Philosophiques* 4, no. 1 (2010b): 549–56.

Carraud, V. "Leibniz lecteur des Pensées de Pascal." *XVIIe siècle* 2 (1986): 107–24.

Carraud, V. *Pascal et la philosophie*. Paris: Presses Universitaires de France, 1992.

Carvallo, S. *Stahl-Leibniz: La controverse sur la vie, l'organisme, et le mixte*. Paris: J. Vrin, 2004.

Chazerans, J-F. "La substance composée chez Leibniz." *Revue Philosophique de France et de l'étranger* 1 (1991): 47–66.

Coudert, A. P. *Leibniz and the Kabbalah*. Dordrecht: Kluwer Academic, 1995.

Couturat, L. *La logique de Leibniz*. Hildesheim: Olms, 1961.

Couturat, L. *De l'infini mathematique*. Paris: Librairie Scientifique et Technique Albert Blanchard, 1973.

Cover, J. A., and J. O'Leary-Hawthorne. *Substance and Individuation in Leibniz*. Cambridge: Cambridge University Press, 1999.

Crockett, T. "Continuity in Leibniz's Mature Metaphysics." *Philosophical Studies* 94 (1999): 119–38.

Duchesneau, F. *Les modèles du vivant de Descartes à Leibniz*. Paris: J. Vrin, 1998.

Duchesneau, F. *Leibniz, le vivant et l'organisme*. Paris: J. Vrin, 2010.

Fichant, M. "L'ontologie leibnizienne de l'action: *actiones sunt suppositorum*." *Philosophie* 53 (1997): 135–48.

Fichant, M. *Science et métaphysique dans Descartes et Leibniz*. Paris: Presses Universitaires de France, 1998.

Fichant, M. "Leibniz et les machines de la nature." *Studia leibnitiana* 35 (2003): 1–28.

Fichant, M. "L'invention métaphysique." Introduction. *G. W. Leibniz: Discours de métaphysique suivi de Monadologie et autres textes*, edited by M. Fichant, 7–147. Paris: Gallimard, 2004.

Fleming, N. "On Leibniz on Subject and Substance." In *Gottfried Wilhelm Leibniz: Critical Assessment*, edited by R. S. Woolhouse, 2:105–27. London and New York: Routledge, 1994. [Reprinted from *Philosophical Review* 96 (1987): 69–95]

Fox-Keller, E. *Making Sense of Life: Explaining Biological Development with Models, Metaphors, and Machines*. Cambridge, MA: Harvard University Press, 2002.

Garber, D. "Leibniz and the Foundations of Physics: The Middle Years." In *The Natural Philosophy of Leibniz*, edited by K. Okruhlik and J. R. Brown, 27–130. Dordrecht: Reidel, 1985.

Garber, D. "Leibniz, Physics and Philosophy." In *The Cambridge Companion to Leibniz*, edited by N. Jolley, 270–352. Cambridge: Cambridge University Press, 1995.

Garber, D. "Dead Force, Infinitesimals, and the Mathematicization of Nature." In *Infinitesimal Differences: Controversies Between Leibniz and His Contemporaries*, edited by U. Goldenbaum and D. Jesseph, 281–306. Berlin and New York: Walter de Gruyter, 2008.

Garber, D. *Leibniz—Body, Substance, Monad*. Oxford and New York: Oxford University Press, 2009.

Garber, D. "Monads on My Mind." In *Leibniz's Metaphysics and Adoption of Substantial Forms*, edited by A. Nita, 161–76. Dordrecht: Springer, 2015.

Gerhardt, C. I. "Leibniz und Pascal." In *Sitzungsberichte der Königlichen Akademie der Wissenschaften zu Berlin*, vol. 2. Berlin: Verlag der Königlichen Akademie der Wissenschaften, 1891.

Gilson, É. "L'infinité divin chez saint Augustin." In *Augustinus Magister: congrès international augustinien*. Vol. I, 569–574). Paris: Études Augustiniennes, 1954.

Goldenbaum, U. "Leibniz as a Lutheran." In *Leibniz, Mysticism, and Religion*, edited by A. Coudert, R. H. Popkin, and G. M. Weiner, 169–92. Dordrecht: Kluwer Academic, 1998.

Goldenbaum, U. "Indivisibilia Vera—How Leibniz Came to Love Mathematics, Appendix: Leibniz Marginalia in Hobbes' *Opera philosophica* and *De corpore*." In *Infinitesimal Differences: Controversies Between Leibniz and His Contemporaries*, edited by U. Goldenbaum and D. Jesseph, 53–94. Berlin and New York: Walter de Gruyter, 2008.

Goldenbaum, U., and D. Jesseph, eds. *Infinitesimal Differences: Controversies Between Leibniz and His Contemporaries*. Berlin and New York: Walter de Gruyter, 2008.

Gueroult, M. *Spinoza I: Dieu*. Paris: Aubier-Montaigne, 1968.

Guitton, J. *Pascal et Leibniz*. Paris: Aubier 1951.

Harera, Rabbai A. C. *Beit Elohim Vesharey Shamaim [The House of God and the Gate to Heaven]*, edited and translated by Y. Nissim. Jerusalem: Yad Ben Ztvi, 2002.

Hatfield, G. "Descartes' Physiology and Its Relation to His Psychology." In *The Cambridge Companion to Descartes*, edited by J. Cottingham, 335–70. Cambridge: Cambridge University Press, 1992.

Ishiguro, H. *Leibniz's Philosophy of Logic and Language*, 2nd ed. Cambridge: Cambridge University Press, 1990.

Ishiguro, H. "Unity Without Simplicity." *The Monist* 81 (1998): 534–52.

Jacob, F. *La logique du vivant*. Paris: Gallimard, 1970.

Jesseph, D. M. "Leibniz on the Foundation of the Calculus: The Question of the Reality of Infinitesimal Magnitudes." *Perspective on Science* 6 (1998): 6–40.

Jesseph, D. M. "Truth in Fiction: Origins and Consequences of Leibniz's Doctrine of Infinitesimal Magnitude." In *Infinitesimal Differences: Controversies Between Leibniz and His Contemporaries*, edited by U. Goldenbaum and D. Jesseph, 215–34. Berlin and New York: Walter de Gruyter, 2008.

Jorati, J. *Leibniz on Causation and Agency*. Cambridge: Cambridge University Press, 2017.

Knobloch, E. "The Infinite in Leibniz's Mathematics: The Historiographical Method of Comprehension in Context." In *Trends in the Historiography of Science*, edited by K. Gavroglu, J. Christianidis, and E. Nicolaidis, 265–78. Dordrecht: Kluwer Academic, 1994.

Knobloch, E. "Galileo and Leibniz: Different Approaches to Infinity." *Archive for the History of the Exact Sciences* 54 (1999): 87–99.

Koyré, A. *From the Closed World to the Infinite Universe*. Baltimore, MD: Johns Hopkins University Press, 1957.

Kretzmann, N., A. Kenny, and J. Pinborg, eds. *The Cambridge History of Later Medieval Philosophy*. New York: Cambridge University Press, 1982.

Kulstad, M. "Leibniz's Conception of Expression." *Studia Leibnitiana* 9 (1977): 55–76.

Kulstad, M. "Some Difficulties in Leibniz's Definition of Perception." In *Leibniz: Critical and Interpretive Essays*, edited by M. Hooker, 65–78. Minneapolis: University of Minnesota Press, 1982.

Lærke, M. *Leibniz lecteur de Spinoza. La genése d'une opposition complexe*. Paris: Honoré Champion, 2008.

Lærke, M. "Spinoza's Monism? What Monism?" In *Spinoza on Monism*, edited by P. Goff, 244–61. Hampshire: Palgrave Macmillan, 2012.

Lærke, M. *Les Lumières de Leibniz. Controverses avec Huet, Bayle, Regis, et More*. Paris: Classiques Garnier, 2015.

Levey, S. "Leibniz on Mathematics and the Actually Infinite Division of Matter." *Philosophical Review* 107 (1998): 49–96.

Levey, S. "Leibniz's Constructivism and Infinitely Folded Matter." In *New Essays on the Rationalists*, edited by R. Gennaro and C. Huenemann, 134–62. New York: Oxford University Press, 1999.

Levey, S. "On Unity: Leibniz- Arnauld Revisited." *Philosophical Topics* 3 (2003): 245–75.

Levey, S. "On Unity and Simple Substance in Leibniz." *Leibniz Review* 17 (2007): 64–66.

Levey, S. "Archimedes, Infinitesimals and the Law of Continuity: On Leibniz's Fictionalism." In *Infinitesimal Differences: Controversies Between Leibiz and His Contemporaries*, edited by U. Goldenbaum and D. Jesseph, 107–34. Berlin and New York: Walter de Gruyter, 2008.

Levey, S. "On Unity, Borrowed Reality and Multitude in Leibniz." *Leibniz Review* 22 (2012): 97–134.

Lodge, P. "Leibniz's Notion of an Aggregate." *British Journal for the History of Philosophy* 9, no. 3 (2001): 467–86.

Lodge, P. "Leibniz's Close Encounter with Cartesianism in the Correspondence with De Volder." In *Leibniz and His Correspondents*, edited by P. Lodge, 162–92. New York: Cambridge University Press, 2004.

Look, B. "On Monadic Domination in Leibniz's Metaphysics." *British Journal for the History of Philosophy* 10, no. 3 (2002): 379–99.

Look, B. "On Substance and Relations in Leibniz's correspondence with Des Bosses." In *Leibniz and His Correspondents*, edited by P. Lodge, 238–61. New York: Cambridge University Press, 2004.

Look, B., and D. Rutherford. Introduction. In *The Leibniz- Des Bosses Correspondence*, edited by B. Look. and D. Rutherford, xix–lxxii. New Haven, CT: Yale University Press, 2007.

Marras, C. "Mirrors That Mirror Each Other." In *Papers from the VIII Internationaler Leibniz-Kongress*, edited by H. Breger, J. Herbest, and S. Erdner, 556–64. Hannover: n.p., 2006.

McDonough, J. K. Review of Smith (2011). *Notre Dame Philosophical Reviews* 4, no. 14 (2012). http://ndpr.nd.edu/news/30317-divine-machines-leibniz-and-the-sciences-of-life-2/.

McDonough, J. K. "Leibniz's Conciliatory Account of Substance." *Philosophers' Imprint* 3, no. 6 (2013): 1–23.

McDonough, J. K. and Nguyen, T. "Monad." In *Routledge Encyclopedia of Philosophy*. Taylor and Francis, 2017. https://www.rep.routledge.com/articles/thematic/monad/v-1.

McKenna, A. *De Pascal à Voltaire. Le rôle des Pensées de Pascal dans l'histoire des idées entre 1670 et 1734*, 2 vols. Oxford: Voltaire Foundation, 1990.

McRae, R. "The Theory of Knowledge." In *The Cambridge Companion to Leibniz*, edited by N. Jolley, 176–98. Cambridge: Cambridge University Press, 1995.

Mercer, C. *Leibniz's Metaphysics: Its Origin and Development*. Cambridge: Cambridge University Press, 2001.

Mesnard, J. "Leibniz et les papiers de Pascal." *Studia Leibnitiana Supplementa* 17 (1978): 45–58.

Mugnai, M. "Leibniz's Nominalism and the Reality of Ideas in the Mind of God." In *Mathesis rationis*, edited by A. Heinekamp, W. Lenzen, and M. Schneider, 153–67. Munich: Nodus, 1990.

Mugnai, M. "Leibniz's Theory of Relations." *Studia Leibnitiana*, Supplement 28.

Murdoch, J. E. "Infinity and Continuity." In *The Cambridge History of Later Medieval Philosophy*, edited by N. Kretzmann, A. Kenny, and J. Pinborg, 564–92. New York: Cambridge University Press, 1992.

Nachtomy, O. "Leibniz on the Greatest Number and the Greatest Being." *Leibniz Review* 15 (2005): 49–66.

Nachtomy, O. *Possibility, Agency, and Individuality in Leibniz's Metaphysics*. New Synthese Historical Library. Dordrecht: Springer, 2007a.

Nachtomy, O. "Leibniz on Nested Individuals." *British Journal for the History of Philosophy* 15, no. 4 (2007b): 709–28.

Nachtomy, O. "Leibniz on Artificial and Natural Machines." In *Machines of Nature and Corporeal Substances in Leibniz*, edited by J. E. H. Smith and O. Nachtomy, 61–80. New Synthese Historical Library. Dordrecht: Springer, 2010.

Nachtomy, O. "A Tale of Two Thinkers, One Meeting, and Three Degrees of Infinity: Leibniz and Spinoza in 1675–78." *British Journal for the History of Philosophy* 19 no. 5 (2011): 935–61.

Nachtomy, O. "Infinity and Life: The Role of Infinity in Leibniz's Theory of Living Beings." In *The Life Sciences in Early Modern Philosophy*, edited by O. Nachtomy and J. E. H. Smith, 9–28. New York: Oxford University Press, 2014.

Nachtomy, O. "Infinite and Limited: Remarks on Leibniz's View of Created Beings." *Leibniz Review* 26 (2016): 179–96.

Nachtomy, O. "Modal Adventures Between Leibniz and Kant." In *The Actual and the Possible*, edited by M. Sinclair, 64–91. Oxford, UK: Oxford University Press, 2017.

Nachtomy, O., "Monads at the Bottom, Monads at the Top, Monads All Over." *British Journal for History of Philosophy* 26, no. 1 (2018): 197–207.

Nachtomy, O., and T. Levanon. "Oneness and Substance in Leibniz's Middle Years." *Leibniz Review* 25 (2015): 69–91.

Naërt, É. *Leibniz et la querelle du pur amour*. Paris: J. Vrin, 1959.

Naërt, É. "Double infinité chez Pascal et Monade." *Studia Leibnitiana* 17, no. 1 (1985): 44–51.

Pasini, E. *Corpo et funzione cognitivi in Leibniz*. Milan: Franco Angeli, 1996.

Phemister, P. "'All the Time and Everywhere Everything's the Same as Here': The Principle of Uniformity in the Correspondence Between Leibniz and Lady Masham." In *Leibniz and His Correspondents*, edited by P. Lodge, 193–213. New York: Cambridge University Press, 2004.

Phemister, P. *Leibniz and the Natural World Activity: Passivity and Corporeal Substances in Leibniz's Philosophy*. Dordrecht: Springer, 2005.

Phemister, P. *Leibniz and the Environment*. London: Routledge, 2016.

Phemister, P., and J. E. H. Smith. "Leibniz and the Cambridge Platonists and the Debate over Plastic Natures." In *Leibniz and the English-Speaking World*, edited by P. Phemister and S. Brown, 95–110. Dordrecht: Springer, 2007.

Pyle, A. *Malebranche*. London: Routledge, 2003.

Rauzy, J. B. "Quid sit Natura Prius? La conception leibnizienne de l'ordre." *Revue de Métaphysique et de Morale* 1 (1995): 31–48.

Rescher, N. "Leibniz's Conception of Quantity, Number, and Infinity." *Philosophical Review* 64, no. 1 (1955): 108–14.

Rey, A-L., and E. Vengeon, eds. "Nicolas de Cues et G.W. Leibniz: Infini, Expression et Singularité." *Revue de Métaphysique et de Morale* 2 (2011).

Riley, P. *Leibniz' Universal Jurisprudence: Justice as the Charity of the Wise.* Cambridge, MA: Harvard University Press, 1996.

Rodriguez-Pereyra, G. *Leibniz's Principle of Identity of Indiscernibles.* Oxford: Oxford University Press, 2014.

Roger, J. *Les sciences de la vie dans la pensée française au XVIII siècle.* Paris: Armand Colin, 1963.

Russell, B. *A Critical Exposition of the Philosophy of Leibniz.* London: Routledge, 1992.

Rutherford, D. "Leibniz on Infinitesimals and the Reality of Force." In *Infinitesimal Differences: Controversies Between Leibniz and His Contemporaries*, edited by U. Goldenbaum and D. Jesseph, 256–80. Berlin and New York: Walter de Gruyter, 2008.

Schmid, S. "The Intrinsic Directedness of Leibnizian Forces." In *Vorträge des X. Internationalen Leibniz-Kongresses*, edited by W. Li, 5:131–41. Hildesheim: Georg Olms, 2016.

Schweitz, L. F. "On the Continuity of Nature and the Uniqueness of Human Life in G. W. Leibniz." In *The Life Sciences in Early Modern Philosophy*, edited by O. Nachtomy and J. E. H. Smith, 205–21. New York: Oxford University Press, 2014.

Serres, M. *Le système de Leibniz et ses modèles mathématiques*, vol. 2. Paris: Presses Universitaires de France, 1968.

Shechtman, A. "Three Infinities in Early Modern Philosophy." Forthcoming.

Simmons, A. "Changing the Cartesian Mind: Leibniz on Sensation, Representation and Consciousness." *Philosophical Review* 110, no. 1 (2001): 31–75.

Sleigh, R. C. *Leibniz & Arnauld: A Commentary on Their Correspondence.* New Haven, CT, and London: Yale University Press, 1990.

Smith, J. E. H. "Leibniz, Microscopy, and the Metaphysics of Composite Substance." Doctoral dissertation, Columbia University, 2000.

Smith, J. E. H. *Divine Machines: Leibniz's Philosophy of Biology.* Princeton, NJ: Princeton University Press, 2011.

Smith, J. E. H., and O. Nachtomy, eds. *Machines of Nature and Corporeal Substances in Leibniz.* Dordrecht: Springer, 2010.

Wilson, C. *The Invisible World.* Princeton, NJ: Princeton University Press, 1995.

Wilson, C. "Leibniz and the Animalcula." In *Studies in Seventeenth-Century European Philosophy*, edited by M. A. Stewart, 153–75. Oxford: Oxford University Press, 1997.

Wolfe, C. T. "Le mécanique face au vivant." In *L'automate: modèle, machine, merveille*, edited by B. Roukhomovsky, S. Roux, and A. Gaillard, 115–38. Bordeaux: Presses Universitaires de Bordeaux, 2012.

NAME INDEX

Adams, R.M., 33, 56, 68–69, 172–73, 183
Aiton, E., 4–5
Anapolitanos, D.A., 93
Andrault, R., 105–6, 107–8, 113, 183–84, 196, 203
Anselm, 10–11, 42–43, 46, 63
Antognazza, M.R., 12, 25, 27–28, 111, 135–36, 158, 159–61, 168, 169–71, 175–76, 196–97
Aquinas, T., 52–53, 82, 84–85
Ariew, R., 60, 69, 70, 86–87, 88, 171, 188
Aristotle, 7, 20–21, 76, 84–85, 105–6, 171–72
Arnauld, A., 2, 31, 86–88, 107, 108, 111, 118–19, 135–36
Arthur, R.T.W., 8–9, 16, 20, 26–28, 37, 54–55, 59, 67, 92–93, 94, 99, 100, 101–2, 103–4, 112, 120–21, 151, 181–82, 186, 193, 195

Baruzi, J., 134–37, 138, 144–45
Bassler, O.B., 26–27, 56–57
Bayle, P., 95, 173
Belaval, Y., 29–30
Bernoulli, J., 26–27, 58–59, 97, 180, 184–85
Bierling, F.W., 180
Bossuet, J-B., 113
Bourguet, L., 84, 85, 92
Brown, G., 57
Burnett, T., 121–22, 136–37
Buzon, F., de., 135–41, 152

Carraud, V., 134–35, 136–37, 148
Chazerans, J-F., 126, 127
Clarke, S., 112, 115, 151–52, 188, 189, 190, 193, 196–97
Clerselier, C., 55–56
Conring, H., 47–48
Couturat, L., 29–30

Crockett, T., 93
Cudworth, R., 193, 201
Curley, E., 63, 66

De Volder, B., 30–31, 32, 33, 92, 94–96, 137, 155, 165, 172, 180, 186–87
Des Bosses, B., 17, 25, 26, 72–73, 82–83, 90–91, 96, 125, 128, 158, 159, 165, 186–87
Descartes, R., 1–2, 7, 10–11, 20, 22, 43–44, 52–60, 100–1, 113–18, 143–44, 168, 172, 174, 201–2
Duchesneau, F., 5, 181–82, 186, 190, 194

Elizabeth, Countess, 40, 47

Fardella, M., 9, 80, 88, 106
Fichant, M., 31–32, 86, 98, 111, 113, 120, 121, 177–78, 184, 185
Fleming, N., 96

Gackenholtz, A.C., 198–99
Galileo, G., 7, 14–16, 35–39, 81, 193
Garber, D., 82, 104, 171, 175–76, 178–79, 180
Gerhardt, C.I., 134–35
Goldenbaum, U., 54–55
Graevious, 135–36
Grew, N., 193, 201
Gueroult, M., 70, 78
Guitton, J., 134–35

Harera, Rabbai, A.C., 84–85
Hatfield, G., 116
Hobbes, T., 4, 54–55, 84–85

Ishiguro, H., 81

Jesseph, D.M., 54–55
Jorati, J., 30–31, 147–48, 178, 194–95

Kant, I., 131, 132, 200
Knobloch, E., 37, 38
Koyré, A., 3
Kulstad, M., 147–48

Lærke, M., 68–69, 72, 74–75, 82, 85, 148–49
Levey, S., 26–27, 82–83, 88–89, 92–93, 118–19
Lodge, P., 87–88, 118–19
Look, B., 24–25, 87–88, 183

Malebranche, N., 1–2, 15–16, 57, 61, 102, 103–4, 109–10, 140–41, 149–50, 156
Marras, C., 144–45
Masham, Lady, 1–2, 113, 123–24, 126, 151–52
McDonough, J.K., 5–6, 87–88, 178
McKenna, A., 134–35
McRae, R., 83–84
Mercer, C., 2
Mesnard, J., 135–36
Meyer, L., 11, 18–19
More, H., 55
Mugnai, M., 29–30, 75
Murdoch, J.E., 7, 24–25

Nachtomy, O., 1–2, 31–32, 39, 47, 99, 103–4, 111, 112, 118–19, 124, 127–28, 151, 167, 184, 195–96
Naërt, É., 134–35, 138, 148–49

Oldenburg, H., 39–40, 136

Pascal, B., 3–4, 7, 102, 107–8, 134–57, 197–98, 201–2

Pasini, E., 120–21, 130
Phemister, P., 33, 110, 179, 186
Pinborg, J., 7
Pyle, A., 1–2

Rabouin, D., 27–28
Rauzy, J.B., 93–94
Remond, N., 23–24, 103–4, 137, 180–81
Riley, P., 137
Roger, J., 1–2
Russell, B., 85, 171
Rutherford, D., 14–15, 24–25, 87–88, 183, 192

Schmid, S., 194–95
Schweitz, L.F., 149–50
Serres, M., 134–35, 155
Simmons, A., 147
Sleigh, R., C., 87
Smith, J.E.H., 5–6, 12, 186, 188–89, 190, 194, 198
Sophie, C., 6, 9–10, 23, 76, 80, 83, 88, 89, 101–2, 126, 137, 146–47, 153–54, 155, 185, 192–93, 195
Spinoza, B., 4–5, 11–12, 18–19, 63–79, 80–81, 83–85, 91, 150–51, 159, 161–62, 168–70, 202
Stahl, G.E., 114–15, 132, 186, 189–90, 194, 202–3

Wagner, G., 134–35, 147
Wilson, C., 1–2, 3, 4–5
Wolff, C., 101–2, 163–64

Zeno, 20–21

TERMS INDEX

abstraction
  mental, 49–50, 75, 83–84, 90–91
action
  course of, 33, 110, 165–66, 167, 171, 173
  power of, 7, 33, 99, 106–7, 173, 175–76, 194–97
  program of, 7, 110, 127, 156, 167
  source of, 34, 110, 171–72, 174–76, 191, 193
  unique, 171
agency, 13, 29–31, 117, 203
aggregate, 10, 24, 29, 58, 88–92, 96–98, 105, 118–22, 124–25, 127, 132–33, 178, 179–80, 183–85
Animalcula, 3, 5–6, 103–4, 184.
  See also under microscope
animate things, 106, 107–8, 113, 200
  inanimate things, 113, 200
appetite/*appétit*, 33, 114–15, 129–30, 191, 193
appetition, 33, 147, 157, 193
attribute
  affirmative, 44–47, 161, 163–64, 168
  positive, 10–11, 44–47, 50–51, 60, 161, 162–63, 166–67

beings
  as imitations of God
    (*see under* imitation)
  created, 7, 12, 19, 30–31, 34, 79, 156, 160, 172, 175, 197–98
  determinate, 24–26, 158, 159, 165–66, 169–70, 171, 176
  entia rationis, 11–12, 19, 29–30, 49–50, 61–62, 65, 79, 158, 176
  infinite (*see* under infinity)
  intelligent creatures, 51, 166, 167

living, 1–6, 12–13, 19, 99, 106, 107–8, 112, 113, 118–19, 127, 134–35, 137–38, 140–41, 145, 151–53, 157, 161, 182, 183–84, 191, 196, 197–98, 200, 201–2
  most perfect/Ens Perfectissimum
    (*see under* God)
  non-beings, 29–30, 50, 71
  nonliving, 1–2, 99, 107–8, 113, 120–21, 179, 182, 184, 190, 200
  rational, 29–30, 47, 167
    nonrational, 167
  true being, 30–31, 49–50, 61, 73, 89, 90, 99, 100–1, 107–8, 179, 181, 191
body, 5–6, 20–21, 86, 90, 92, 104–5, 114–15, 117, 124–25, 129–31, 172, 175, 180–82, 183–84, 185, 186, 194
  as a natural machine
    (*see under* machine)
  organic bodies as machines to the least of their parts, 1–2, 118, 123, 152, 182, 184, 197–98

change
  principle of, 33, 172–73
classification, 72, 198–200
cognitive capacity
  limited, 7, 148, 167
concept
  complete, 31, 109, 127–28, 166–67, 169–70, 171, 185, 198–99
contingency, 41–42, 167
creatures
  as infinite, 141, 158 (*also see under* infinity)
  finite, 7, 12, 79, 141–42, 156, 159
  intelligent, 166, 167
  properties of, 161, 167

215

divine essence, 156, 158, 165, 176.
     Also see under God
domination
domination relation, 124, 182–85

ego, 96–98, 141, 168–69, 179–80, 185
emboîtement, 5, 126, 132, 184
entelechy, 1–2, 13, 98, 99, 105, 107, 110–11,
     121, 126, 155, 171–72, 177–78, 180,
     181, 186, 190–92, 195, 203
enveloppement, 5, 129, 157
existence, 7, 8–9, 10–11, 23, 29–30, 40,
     42–43, 47, 49, 59, 63–65, 68, 76, 78,
     81, 84–85, 91, 97–98, 100–1, 106,
     127–28, 142–43, 163, 164, 170
     claim for, 164
     nonexistence, 38–39, 41–43

force
     as cause of motion, 173–74
     as derived from a divine source, 174, 175
     derivative, 14, 32, 172–73
     inherent, 34, 174–75, 203
     motive, 172, 192
     primitive, 7, 14, 30–31, 32, 51, 94, 109–10,
          111, 158, 160–61, 171–76, 177–78,
          186–87, 191, 194–96, 200, 202, 203
     source of, 173–74

Galileo's Paradox, 14, 35, 41, 49, 61–62
god/creator
     concept of, 40, 50, 51–52
     divine essence, 158, 159, 165, 176
     divine mind, 166–67
     imitations of, 135, 156, 165, 166–67,
          170, 175–76
     infinity of, 7, 10–11, 17–19, 52, 55, 56, 66,
          141, 142–43, 150–51, 158, 161, 176
     positive attributes of, 10–11, 44, 45, 47,
          161, 162–63, 166
     proof of existence, 46–49, 60, 63–65,
          68–69, 142–43
          Descartes', 10–22, 42–43, 44, 48, 63
          Anselm's, 10–11, 42–73
     the most perfect Being (*Ens
          Perfectissimum*), 7, 9–11, 16–17,
          18–19, 28, 38–39, 41–49, 51–52, 56,
          60, 61, 63–64, 68–69, 72–73, 78, 158,
          162–63, 165, 168, 176
grouping, 76, 84, 198–99

harmony, 30–31, 101–2, 132, 134–35, 137,
     139–41, 146–47, 149–50, 155, 156,
     163, 197–98

identity, 30–31, 101–2, 118, 161, 162–63, 168
imagination, 49–50, 65, 66, 70–71, 83–84,
     191, 192–93
immensum/immense, 18, 55–56, 67–68,
     72–73, 78, 102–3, 145–46
individual
     complete notion of, 12, 30–32, 86,
          111, 177–78
     rational, 167
     substance, 110, 111–12, 127–28, 160, 165,
          171, 177–78, 180–81
indivisibility, 14, 20, 64–65, 72–73, 81–83,
     86, 100, 104, 105–6, 111
     divisibility, 20, 23–24, 57, 82, 86, 100–1,
          104, 107–8, 122–23, 134–35, 146–47,
          150–51, 152, 155, 157
infinity
     absolute, 12, 17, 18–19, 65, 72–73, 133,
          141, 150–51, 156n67, 161
     actual, 2, 7, 9–10, 20–21, 22, 22n12, 23,
          24, 29, 139–40, 147n47, 148–49
     degrees of, 11–12, 11n33, 14, 18, 19n7,
          27–28, 63, 65, 133, 135, 150–51,
          160n4, 160–61, 168–70, 175–76
     indefinite, 5–6, 22, 24n17, 35, 52, 53,
          54–57, 59, 77–78, 86–87, 87n19,
          92, 190
          indefinitely divisible, 54–55
     infinite being, 10–11, 14, 16–17, 29–30,
          31, 35, 38–39, 45, 45n16, 49, 50–51,
          60, 61–62, 64–65, 67, 68–69, 70, 71,
          78, 135, 142–43, 159, 161–63, 197–98
     infinite creator, 7, 12, 79, 112, 122n19,
          151–52, 156
     infinite magnitude, 10–11, 51, 52, 73,
          77n28, 102, 144n39
     infinite number, 8, 10–11, 14–19, 24–30,
          31–32, 35–39, 37n3, 40, 44–45, 46–52,
          53, 57–59, 60–62, 64, 65, 68–69, 71,

73, 78, 106n14, 108, 122, 123, 124,
    127–28, 142–43
infinite perfection, 17, 51, 52
infinite quantity, 25–26, 37, 38, 71
infinite series, 9, 14, 15–16, 26–27,
    28, 29–30, 31–32, 38, 57, 59, 91,
    129, 143–44
infinite substance, 56n35, 64, 70, 72,
    73, 81
Infinitum, 1–2, 2n5, 11–12, 11n33, 18–19,
    19n7, 20, 23–24, 64n5, 67n10, 71–72,
    79, 98n37, 106n14, 107, 112, 113,
    124, 129, 141–42, 151–53, 168–69,
    182, 202–3
  infinitum tantum, 61–62, 65, 168–69
  of attributes, 18n6, 66, 66n8
  of creatures, 2n5, 102, 106, 146n45,
    150–51, 154
  of divine attributes, 169
  of God, 7, 10–11, 17, 18–19, 52, 55, 56, 66,
    141, 142–43, 150–51, 158, 161, 176
  of worlds, 103–4, 145–46, 146n45
  maximum, 11–12, 11n33, 12n34,
    18–19, 19n7, 46n18, 47–48, 58,
    69, 71–72, 78, 79, 158, 161–62,
    168–71, 175–76
    maximum in its own kind, its own
      kind, 12, 12n34, 18–19, 19n7, 39–40,
      65, 79, 158, 170–71, 175–76
  Omnia, 11–12, 11n33, 18–19, 19n7,
    39–40, 45–46n17, 65, 68–69, 71–73,
    77n25, 104–5, 168–69, 169n15
    omnia sui generis, 65
  potential, 9–10, 14, 20, 24, 30
  quantitative, 156n67
    non-quantitative, 52
  senses of, 16–17, 35, 50–51, 52, 56–57,
    61–62, 78–79
  syncategorematic, 125
  types of, 65
    finite, 3, 5–6, 7, 8–9, 12, 15, 25–27,
      29, 34, 36–37, 38, 53, 56–57, 64, 66,
      70–71, 77–78, 80, 91, 122–23, 141–44,
      146–47, 148, 149–51, 156, 159–60,
      172, 173–74, 175, 183, 189
intelligence, 12, 155, 192
  intelligent creatures, 51, 166, 167

intelligibility, 28–29, 43, 45, 46
  affirmative, 164

judgment, 75, 83–84, 131, 167

life, 1, 23, 30, 34, 99, 115–16, 122, 127,
    129, 151–52, 153–54, 157, 174, 179,
    182–83, 186
  living beings/Non-living beings
    (*see under* being)
  living mirror (*miroir vivant*), 2–3, 106–7,
    129–30, 134, 171
  phenomena of, 115–16, 127, 188–89, 190,
    194, 196, 197, 198, 201–3
  principle of, 7, 105–6, 129, 174,
    191, 194
  source of, 23, 30, 190–91, 203
life-sciences, 3–5, 7, 12, 190–91, 197
logical space, 171
  as the space of all possible
    things, 166–67

machines
  animated, 107, 128–29
  artificial, 1–2, 5–6, 99, 108, 112, 113, 138,
    151–52, 182, 188–89, 202
  divine, 5, 6, 112, 120, 131, 151–52,
    181–82, 186, 188–89,
    197–98, 202–3
  hierarchical structure of, 182
  natural, 1–2, 7, 99, 112, 113, 135, 139,
    151–53, 155, 181–84, 186, 188–89,
    190, 196–97, 200, 202–3
  nested, 124
  organic bodies as machines to the least of
    their parts, 1–2, 118, 123, 152, 182,
    184, 197–98
  secte machinale, 154
  maximum, 2n7, 11–12, 11n33, 12n34, 18–19,
    39–40, 46n18, 47–48, 58, 65, 69,
    71–72, 78, 79, 158, 161–62, 168–71,
    169n15, 175–76
  greatest figure, 162
  greatest number (*see under* number)
  in its kind, 79, 170–71
  maximal knowledge, 162
  maximal power, 162

mechanical philosophy, 190
mechanism, 5–6, 120, 124, 132, 172, 188, 192, 196–97, 200, 201–3
microscope, 5–6, 103–4, 154, 182
  microscopic animals,/*Animalcula*, 104, 140–41, 184
  microscopic discoveries, 184
  microscopical observations, 6, 103–4, 184
  microscopists, 4–6, 184
mirror. *See* living mirror
mite (ciron), 3–4, 103–4, 134
monad, 2–3, 23, 30–31, 32, 89–90, 91–92, 95, 98, 112, 125, 129–30, 137, 139–41, 145, 146–47, 155, 157, 165, 177, 192, 194–95
  as true unit, 140–41, 155, 177
  dominant, 98, 177–78, 179, 183–84, 185, 187
  hierarchy of, 98, 177
  simple, 178, 180, 181
monism, 82
morality, 13, 37, 166, 202–3
  moral imperfection, 166
multitude argument, 23, 28, 88–89, 178

nature
  reenchantment of, 13, 201
neoplatonism, 158, 160, 175–76
nestedness, 124, 132, 184
  Nested structure, 2, 5–6, 112, 113, 124, 127, 129, 130, 146–47, 151–53, 184, 202–3
notion
  impossible, 39–40, 41–43, 46, 49, 58
  possible, 49
number
  greatest, 39–40, 42, 44, 46–50, 61–62
  infinite (*see under* infinity)
  numerical ascriptions, 74, 75, 77–78, 83–84
  of all numbers, 15–16, 37, 39–41, 48–50, 57, 162
  unbounded (interminatum), 77–78

omnipotence, 106–7, 162, 175. *Also see* power: absolute

one
  as the foundation of numbers, 86
  in number, 80
  oneness, 80, 85, 86–88, 90–91, 94, 96
  uniqueness, 64–65, 73, 76, 78, 81, 82, 149–50, 166, 171
  unum, 15, 64, 82–83, 84–85, 86, 104–5
    unum per accidens, 82
    unum per se, 82, 88
organism, 1–2, 113, 124, 128, 132, 188, 200, 202–3
organon, 130

paradox
  Zeno's, 20–21
part
  and whole, 17, 18
passivity, 161, 165
perception, 31n30, 33, 91–92, 95, 96, 97, 114–15n4, 129–30, 134–35, 136, 142–43, 145, 146–48, 153, 155, 160, 165, 170, 183n13, 186, 193, 194–96
perfection
  absolute, 7, 18, 71, 72–73, 166, 168
  and infinity, 7, 8–9, 78, 160, 162–63, 165, 168, 170–71
  and reality, 161
  degrees of, 161, 164, 165, 168, 169–70, 175–76, 183
  God's, 10n31, 60n42, 61–62, 162–63
  in kind, 162–63
  limited, 160
  most perfect being, 7–9, 10–11, 10n31, 16–17, 18, 19, 38–39, 40, 42–52, 56, 60, 61, 63–64, 63–64n2, 68–69, 72–73, 78, 158, 162–63, 165, 176
  metaphysical, 166, 167
  moral, 166, 167
  most perfect world, 163
  of the world, 132, 164
  of individuals, 164
  the absolute sense of, 162–63
plastic natures, 193, 201–2
pluralism, 91
  methodological, 198, 199–200
point of view (Unique/Limited), 129–30, 154, 167

possibility
  genuine, 41–42
  logical, 64–65
  possible things, 47, 166–67
    logical space of, 166–67
  possible worlds (*see under* world)
power
  absolute, 34, 175
  as appetition, 33, 191, 194–95
  entelechy (*see under* entelechy)
  of action, 7, 33, 106–7, 109, 111–12, 173–74, 175–76, 194–97
predicates
  unique structure of, 127–28, 166
prescription, 167
principle of the best, 167
privation, 160–61, 170–71
  as imperfection, 166
  privative properties of creatures, 167
  zero as privation, 166–67
program
  of action, 7, 110, 127, 156, 167

real definition, 28, 29–30, 31–32, 43–44, 68
reality, 158, 178–80, 184–85, 187
  degrees of, 78, 161
  limited reality, 164
reflection, 49–50, 51, 141–42, 145, 167
res extensa, 20, 100–1, 115
  extension, 18–19, 23, 53, 55–56, 71–72, 100–1, 105, 107–8, 168, 169, 172, 192, 201
rota Aristotelis, 36, 89–91, 96, 100–1, 107–8, 121, 144

series
  infinite, 9, 14, 15–16, 26–27, 28, 31–32, 38, 57, 59, 91, 129, 143–44
  law of the, 14, 28–32, 57, 82–83, 91, 97, 109, 160–61, 171
soul, 1–2, 13, 23, 95–96, 101–2, 104–8, 115–16, 118, 129, 153–54, 170, 175, 179, 180–82, 183–84, 185, 186–87, 190–93, 194–95, 196–97, 201
  entelechy, 1–2, 13, 99, 105, 107, 110–11, 121, 126, 155, 171–72, 177–78, 180, 181–82, 183–84, 186–87, 190–92, 195, 203

spontaneity, 167
substance
  aggregates of, 96, 119
  collections of, 24n15, 179–80
  complete, 141, 179, 180, 181, 192–93
  composite, 178, 179–80
  corporeal, 2n5, 33n39, 107, 171–72, 175, 178, 179n5, 180n6, 180n7, 186, 187, 192
  created, 7, 12, 12n34, 13, 21, 27–28, 34, 73n21, 79, 106–7, 110, 156, 158, 160–61, 165, 171n19, 175–76, 194–95, 197–98
  individual, 9, 29–32, 32n36, 33, 109, 127–41, 156, 160–65, 171, 172, 173, 177–78, 180
  indivisible, 66–67
  indivisibility of, 88, 89, 90
  infinite, 56n35, 64, 70, 72, 73, 81
  simple, 23, 31n30, 91, 92n26, 147, 170, 178–79, 180n6, 195
  unique, 81, 157, 195
  syncategorematic approach, 14, 16, 17, 24, 38, 54–55, 56–57, 61, 65, 71, 77–78, 125, 133, 143–44, 158, 169

teleology, 21, 94, 120–21, 132, 188–89, 190, 197–98, 201–2
telos, 121, 132, 133

unity
  arithmetical, 90–91
  by aggregation, 82, 87–88
  mathematical, 93
  metaphysical, 82–83, 90, 91
  substantial, 23, 82, 86–89, 96, 107–8, 111, 118–19, 121–22, 157
  true, 17, 58, 72–73, 86, 100–1, 105, 107, 118–19, 121, 125, 141, 170

world
  most perfect/Best, 163–64
  possible worlds, 8, 9, 110, 163–64, 202
  he created/actual, 9, 149–51, 156, 164, 183, 202

Zeno's paradoxes. *See under* paradox

www.ingramcontent.com/pod-product-compliance
Ingram Content Group UK Ltd.
Pitfield, Milton Keynes, MK11 3LW, UK
UKHW022153230426
12049UKWH00003BA/79